AF556791

SELECTED MOLECULAR REARRANGEMENTS

SELECTED MOLECULAR REARRANGEMENTS

T. S. STEVENS, F.R.S.
Professor of Chemistry
University of Strathclyde, Glasgow

W. E. WATTS
Senior Lecturer in Chemistry
New University of Ulster, Coleraine

VAN NOSTRAND REINHOLD COMPANY
LONDON
NEW YORK CINCINNATI TORONTO MELBOURNE

VAN NOSTRAND REINHOLD COMPANY LIMITED
25–28 Buckingham Gate, London, SW1E 6LQ

INTERNATIONAL OFFICES
New York Cincinnati Toronto Melbourne

Library of Congress Catalog No.: 72–5927

ISBN 0 442 07983 4

First published 1973

Printed in Great Britain by
William Clowes & Sons, Limited
London, Beccles and Colchester

Contents

Foreword

Molecular rearrangements have provided many of the keys which have opened the doors on our understanding of organic chemistry. At first they presented problems which appeared to undermine the very principles on which the theory of organic chemistry was built. However, in elucidating these problems the organic chemists have been led to a much deeper understanding of their subject. The pinacol rearrangement and the Wagner–Meerwein shift led to the acceptance of carbonium ions as intermediates and played an important part in the development of our understanding of substitution reactions. The Beckmann rearrangement paved the way to the stereochemistry of the oximes and thereby to the understanding of many other stereochemical problems. The cationic rearrangements undergone by the terpenoids led chemists to become interested in biosynthetic sequences, and it is not too much to say that the recent upsurge in the study of biosynthesis finds its origin in the rearrangements of the monoterpenes. Very much more recently, rationalisation of electrocyclic rearrangements paved the way to the concepts of orbital symmetry.

Seldom have the pioneers who first discovered new fields of chemistry been willing or able to write about their discoveries for the benefit of the non-specialist audience and for students. T. S. Stevens, by discovering the rearrangement which bears his name, opened up a field which as the present text shows is still only part surveyed and some of which is yet veiled in mystery. Together with Dr. Watts he has written a text, which has that rare quality for an undergraduate text, of complete originality. The book is certain to be enjoyed by students and teachers alike, for many years to come. At present it is an up-to-date monograph; in the future it will be a classic for its economy of words and its clarity of exposition. The book is to be read not just to learn about molecular rearrangements, but to gain new understanding of organic chemistry in all its many facets.

St. Andrews
March, 1972

J. M. TEDDER

Preface

Sixty years ago the corpus of organic chemistry included a number of reactions involving skeletal rearrangement, as intriguing and isolated anomalies. These have increased in number and diversity and, with growing understanding, in significance. In wide fields there is an extensive and expanding appreciation of mechanism; other cases are still mockingly enigmatic. Few organic chemical investigations can be free from their occasional incursion, at worst as a bewildering surprise, with fuller comprehension as an informative part of the picture. Finally, these transformations often take place smoothly and predictably, so that they can constitute useful synthetic methods which on occasion provide easy access to bizarre structures.

As the title implies, we have not attempted to include all recorded rearrangements, but most of the well-known types are discussed. It is intended that the book should offer good background reading for a bright undergraduate or a post-graduate student, and provide a substantial preliminary introduction for a research worker commencing an investigation in this field. We have attempted to present the subject as a growth point in chemical theory; while it would have been out of place to assemble any full citation of the very extensive literature, we have included sufficient references to recent papers and review articles to enable a reader to embark on his own more comprehensive study of the literature.

We would like to thank Professor Lord Tedder, University of St. Andrews, who read the original manuscript of this book and suggested many ways in which it could be improved, the staff of the Van Nostrand Reinhold Company for expert editorial assistance, and Mrs.

I. M. McLean and her staff, New University of Ulster, for efficient and painstaking typing services.

Glasgow and Coleraine
March, 1972

T. S. Stevens
W. E. Watts

1

Introduction

Never confuse the unusual and the impossible

RUPERT PSMITH

Credo quia impossibile

After TERTULLIAN

An Early Embarrassment

Molecular rearrangements have attracted the attention of three generations of organic chemists, in the first instance because they constitute a distressing, if not scandalous, crack in the foundations of structural chemistry. The structure theory erected by Kekulé and Couper was based on three indispensable postulates:

(a) a one-to-one correspondence between structural formulae, built according to the rules, and isolable compounds; and less precisely:

(b) similar atomic groupings are associated with similar reactions;

(c) the structure(s) of the product(s) of a reaction bears a rational relation to the structure(s) of the initial material(s).

This last is Hückel's '*principle of minimum structural change*', to which skeletal rearrangements are repugnant.

Liebig discovered the benzilic acid change in 1838, before the advent of the structure theory. In 1860, Fittig observed the pinacol-pinacolone rearrangement, which was recognized by Butlerov in the next decade as involving a change in the carbon skeleton.

There were three main possibilities for congenially assimilating these embarrassing findings.

(a) The change could be regarded as the intramolecular analogue of a known intermolecular process. For example, reduction of *o*-nitrophenyl benzoate (1) gives *o*-hydroxybenzanilide (2).

$$\text{(1) } o\text{-}C_6H_4(OCOPh)(NO_2) \longrightarrow [o\text{-}C_6H_4(OCOPh)(NH_2)] \longrightarrow o\text{-}C_6H_4(OH)(NHCOPh) \text{ (2)}$$

The intermediate ester spontaneously undergoes a (structurally convenient) aminolysis, analogous to the reaction of phenyl benzoate with aniline (i.e., $PhCOOPh + NH_2Ph \rightarrow PhCONHPh + HOPh$). This reassuring situation is not general. Thus, the intermolecular equivalent of the pinacol change:

$$Me_2C(OH){-}CMe_2(OH) \xrightarrow[H_2SO_4]{\text{dil.}} MeC({=}O){-}CMe_3$$

would be

$$Me_3COH + HOCMe_3 \xrightarrow[H_2SO_4]{\text{dil.}} Me_2CO + CMe_4$$

which has not yet been observed!

(b) The migrating entity might separate from the rest of the molecule to become kinetically free, and then re-attach itself in the new position. This appears to be true for the Orton rearrangement of *N*-chloroacetanilide:

$$C_6H_5{-}N(Cl)(Ac) \xrightarrow{HCl} C_6H_5{-}NHAc + Cl_2 \xrightarrow{-HCl} Cl{-}C_6H_4{-}NHAc$$

in which the intermediate formation of elemental chlorine is well established (see Chapter 9).

The validity of this type of mechanism can often be tested by a 'mixture experiment'. In this, the rearrangements R—A—B → A—B—R and R′—A′—B′ → A′—B′—R′ are carried out together; A′, B′, R′ are substituted variants of A, B, R, or, better, isotopically labelled forms. Then the intermolecular mechanism should be disclosed by production of the 'scrambled' products, A—B—R′ and A′—B′—R. It will be readily understood that the rates of the two individual rearrangements must not be so disparate that one is complete before the other is well started.

(c) As the other extreme, it could be suggested that the migrating group becomes attached to the destination atom before leaving the point of departure, i.e., we may postulate a cyclic intermediate. There are indeed cases, like the isomerisation of (3), for which this is valid.

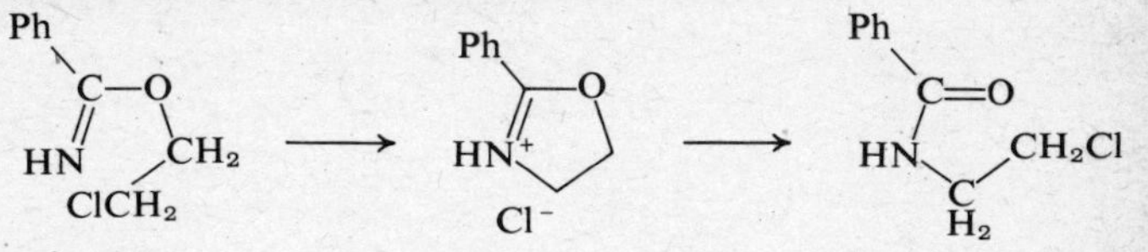

(3)
(liquid; no Cl^-) → (solid; positive test for Cl^-) → (liquid; no Cl^-)

Erlenmeyer proposed the following sequence for the pinacol rearrangement:

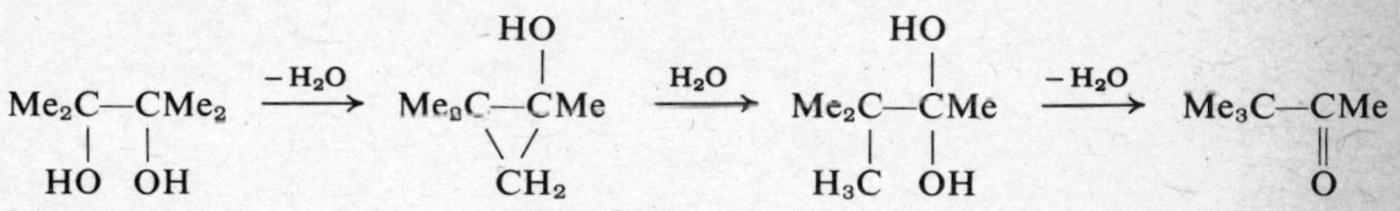

$$\underset{\text{HO}\ \ \text{OH}}{Me_2C{-}CMe_2} \xrightarrow{-H_2O} Me_2C{-}C(OH)Me\ (CH_2\ \text{bridge}) \xrightarrow{H_2O} Me_2C(CH_3){-}C(OH)Me\ (OH) \xrightarrow{-H_2O} Me_3C{-}C(=O)Me$$

To fit this theory to a tetraarylethane-1,2-diol, it was argued that the reaction would proceed via such a structure as (4), so that the point of attachment of the migrating aryl group changes from C_1 to C_2. Montagne, however, showed that, in the rearrangement of tetra-(*p*-chlorophenyl)ethane-1,2-diol, all four groups remain *p*-substituted. This result does not exclude the possibility that a *three-membered* ring intermediate may be formed in reactions involving 1,2-migration of aryl groups (see Chapter 2; phenonium ions).

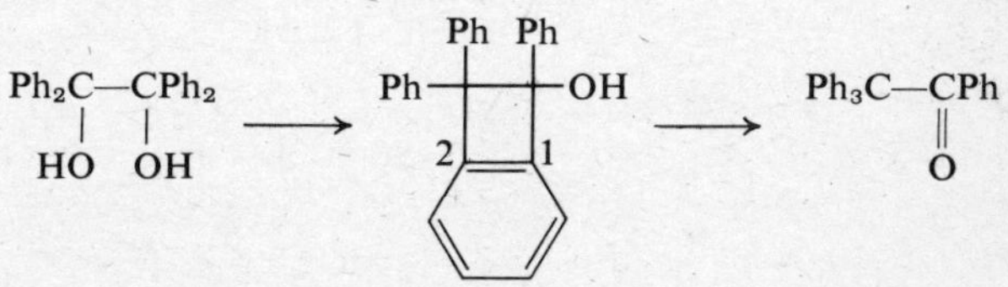

$$Ph_2C(OH){-}CPh_2(OH) \longrightarrow (4) \longrightarrow Ph_3C{-}C(=O)Ph$$

(4)

Somewhat later, there was strong support for a similar explanation of the interconversion of camphene hydrochloride (5) with the epimeric bornyl and isobornyl chlorides (6).

It was postulated that the rearrangement proceeded by way of tricyclene (7). This suggestion was conclusively refuted in two ways.

First, tricyclene proves to be stable under conditions that lead to the rearrangement. Secondly, the optically active halides can be interconverted without racemisation—tricyclene has a plane of symmetry.

(5) ⟶ (6) (7)

Growing Light on 1,2-Shifts

Up to this point, there was no effective attack on the problem on a broad front. Following the work of Stieglitz and others, which was very substantial though motivated by pre-Lewis electronic theories, in 1932 Whitmore achieved an extensive generalisation [1]. Consider the system:

$$R-\underset{R}{\overset{R}{A}}-\underset{R}{\overset{R}{B}}-X \xrightarrow{-X^-} R-\underset{R}{\overset{R}{A}}-\underset{R}{\overset{R}{B^+}}$$

If X splits off as an anion, then when A = B = carbon, the product is a carbonium ion in which B carries an electron sextet. Stabilisation can follow in several ways.

(a) Another anion (Y^-) can link to B, completing a simple substitution reaction.

(b) If one group R is H, a proton may split off, in combination with X^- or otherwise, giving an alkene, $R_2A{=}BR_2$.

(c) If one group R is H, it may migrate to atom B, with possible re-attachment of X^- to A, as in the easy interconversion of isobutyl and t-butyl bromides:

$$Me-\underset{H}{\overset{Me}{C}}-\underset{H}{\overset{H}{C}}-Br \longrightarrow Me-\underset{H}{\overset{Me}{C}}-\underset{H}{\overset{H}{C^+}} \longrightarrow Me-\overset{Me}{\overset{+}{C}}-\underset{H}{\overset{H}{C}}-H \longrightarrow Me-\underset{Br}{\overset{Me}{C}}-\underset{H}{\overset{H}{C}}-H$$

(d) Finally, a group R may migrate in the same way:

$$Me_3CCH_2OH \xrightarrow{HBr} Me_3C\overset{+}{C}H_2 \longrightarrow Me_2\overset{+}{C}CH_2Me \xrightarrow{Br^-} Me_2CBrCH_2Me$$

$$Me_2\overset{+}{C}CH_2Me \xrightarrow{-H^+} Me_2C{=}CHMe$$

Wagner–Meerwein rearrangements, similar in type, led to great confusion in early structural work on monoterpenes. Note that the second carbonium ion shown above, having a tertiary carbonium carbon, is more stable than the first, primary one. This gain in stability was regarded as providing the driving force for the reaction.

Scheme 1

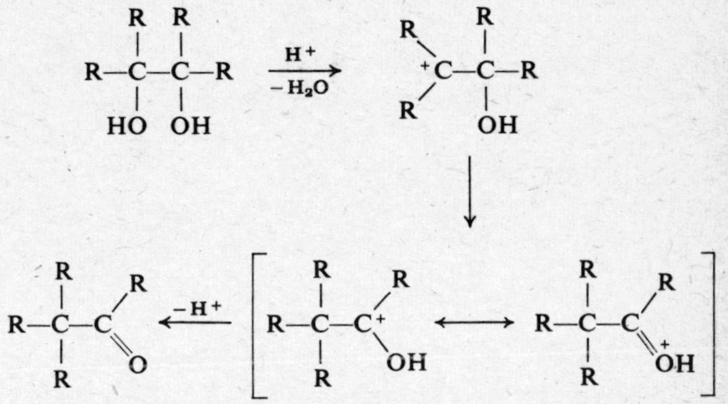

$$R_2C(OH){-}C(OH)R_2 \xrightarrow[-H_2O]{H^+} R_2\overset{+}{C}{-}C(OH)R_2$$

$$\downarrow$$

$$R_3C{-}C({=}O)R \xleftarrow{-H^+} \left[R_3C{-}\overset{+}{C}(R){-}OH \longleftrightarrow R_3C{-}C(R){=}\overset{+}{O}H \right]$$

The pinacol-pinacolone change was formulated in a similar way (Scheme 1). Note that resonance stabilisation in the rearranged cation promotes the rearrangement. Similarly, the Beckmann transformation of oximes can be satisfactorily accommodated:

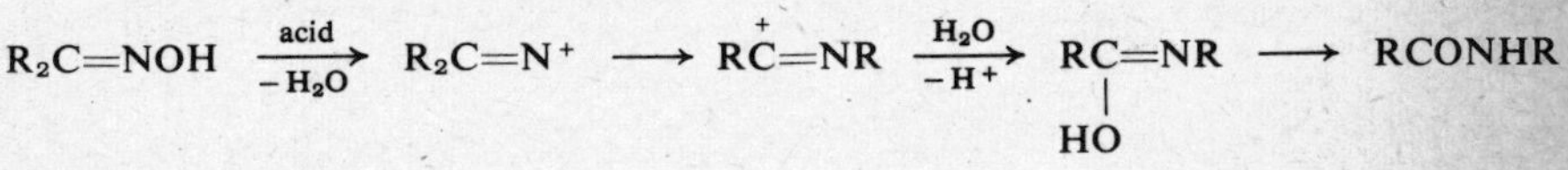

$$R_2C{=}NOH \xrightarrow[-H_2O]{acid} R_2C{=}N^+ \longrightarrow R\overset{+}{C}{=}NR \xrightarrow[-H^+]{H_2O} \underset{HO}{RC}{=}NR \longrightarrow RCONHR$$

The conversions of RCO_2H into RNH_2 by the Hofmann, Curtius, and Lossen rearrangements can all be discussed on similar lines, but postulating as sextet intermediate an uncharged nitrene, $RCO\ddot{\underset{..}{N}}$ (see Chapter 3).

Rearrangements of this kind, in which the destination atom is electron-deficient, are often designated *nucleophilic rearrangements* since the migrating group appears to behave as a nucleophile. It is often stated that the group moves 'with its bonding pair of electrons.' This is true as an expression of accountancy, but can be misleading; it must not be assumed that the group is actually detached *as an anion* before proceeding to its new position.

A counterpart of the nucleophilic rearrangements is provided by a series of *electrophilic rearrangements* in which the destination atom is initially electron-rich, carrying an unshared electron-pair and usually a negative formal charge. Examples are given below.

(a) Rearrangements of amine oxides (Meisenheimer):

$$\overset{-}{O}-\overset{+}{N}MePh(CH_2Ph) \xrightarrow{100^\circ} O-NMePh(CH_2Ph)$$

(b) Rearrangement of quaternary ammonium ions (Stevens):

$$PhCOCH_2\overset{+}{N}Me_2(CH_2Ph) \xrightarrow{\text{base}} PhCO\overset{-}{C}H\overset{+}{N}Me_2(CH_2Ph) \longrightarrow PhCOCH(CH_2Ph)NMe_2$$

(c) Rearrangement of ethers (Wittig):

$$PhCH_2O(CH_2Ph) \xrightarrow{PhLi} Ph\overset{-}{C}HO(CH_2Ph) \longrightarrow PhCH(CH_2Ph)\overset{-}{O} \xrightarrow{H^+} PhCH(CH_2Ph)OH$$

Note that in (b) and (c) it is necessary to use an alkaline reagent of suitable strength to remove a proton from the destination carbon atom.

Rearrangements of free radicals are foreseeable, where the destination atom carries an electron septet in contrast with the sextet for nucleophilic and the partially unshared octet for electrophilic rearrangements. Cases are well established [2], but the topic has been less extensively studied and is not developed in detail in this book. Only aryl groups appear to migrate.

$$\underset{(8)}{Ph_3CCH_2CHO} \xrightarrow{(t\text{-}BuO)_2} Ph_3CCH_2\dot{C}O \xrightarrow{-CO} Ph_3C\dot{C}H_2 \longrightarrow Ph_2\dot{C}CH_2Ph \longrightarrow Ph_2CHCH_2Ph$$

In the case of (8) a t-butoxy radical, from dissociation of the peroxide, abstracts the aldehydic hydrogen atom giving a free acyl radical, and thence, by loss of carbon monoxide, free 2,2,2-triphenylethyl. Migration of phenyl gives the obviously more stable 1,1,2-triphenylethyl which here 'picks up' a hydrogen atom, giving the final product in high yield. In other cases, the rearranged radicals dimerise, disproportionate to substituted alkane and substituted alkene, or lose a hydrogen atom to give the alkene. The situation may be further complicated by polymerisation.

More recent work has thrown considerable light on the nature of the 1,2-nucleophilic and -electrophilic rearrangements. Mixture experiments have indicated that in general the processes are intramolecular. This conclusion is reinforced by the retention of optical asymmetry in systems that include an asymmetric centre which would be racemised in a fission-recombination process; a tricovalent carbon atom, charged or otherwise, cannot long remain chiral. In particular, if the atom of the migrating-group attached to the rest of the molecule is asymmetric, migration takes place with retention of configuration in a nucleophilic rearrangement, and with retention plus more or less racemisation in an electrophilic rearrangement.

If the destination atom is chiral, it appears that a shift of the Whitmore type normally proceeds with inversion at that atom; the rearranging molecule has to assume a conformation such that the migrating-group occupies an *anti*-parallel relationship to the separating X. The process is a concerted one, with a transition state in which the migrating-group has moved towards its destination while the departing anion is in course of receding. This is an example of the frequently encountered phenomenon of neighbouring-group participation.

Stereochemical influences can be decisive in reactions taking place in conformationally rigid polycyclic structures; they define the observed products of reaction, and, in so doing, allow inferences as to mechanism.

Kinetic studies of the effect of substitution on ease of rearrangement have often served to exclude some suggested mechanisms, but may remain in at least rough consonance with more than one further alternative. In particular, many attempts have been made to assign quantitative 'relative migratory aptitudes' to individual groups, either on the basis of reaction rates or by setting pairs of groups in competition on the same atom of departure:

$$\mathrm{RA{-}BR'} \longleftarrow \mathrm{A{-}B}\begin{matrix}\diagup\mathrm{R}\\ \diagdown\mathrm{R'}\end{matrix} \longrightarrow \mathrm{R'A{-}BR}$$

Experience showed that, at best, an observed order of aptitude would apply to only one type of rearrangement, shedding considerable illumination on the restricted field, however. Conformational influences can often disturb such generalisations (see Chapter 2).

We have seen that, in general, the first sextet-bearing entity proposed in the Whitmore scheme, e.g. $R_2C{=}N^+$ for the Beckmann change, does not become kinetically free; the ejection of X is coupled with the migration of R. In the case of electrophilic rearrangements, on the other hand, the lone-pair destination atom belongs to an entity that exists independently in the sequence:

$$\text{H—A—B—R} \xrightarrow{-\text{H}^+} \text{:}\overset{-}{\text{A}}\text{—B—R} \longrightarrow \text{R—A—}\overset{-}{\text{B}}\text{:} \xrightarrow{\text{H}^+} \text{R—A—B—H} \quad \text{(a)}$$

$$\downarrow$$

$$\text{A=B} + \text{:R}^- \longrightarrow \text{A=B} + \text{HR} \quad \text{(b)}$$

Here, group migration (path a) is much less favoured and often encounters a competing and even predominant process of eliminative fission (path b) as shown. It is notable that the suprafacial shift

$$\overset{+}{\text{C}}\text{—C(R)} \longrightarrow \text{(R)C—}\overset{+}{\text{C}}$$

is, by Woodward–Hoffmann rules, an 'allowed' process, whereas the parallel electrophilic shift

$$\overset{-}{\text{C}}\text{—C(R)} \longrightarrow \text{(R)C—}\overset{-}{\text{C}}$$

is forbidden. Accordingly, 1,2-electrophilic shifts are less common than nucleophilic and, when they do occur, they are not concerted one-stage processes (see Chapter 8).

Longer-range Migrations

Many of these reactions have the appearance of intramolecular nucleophilic substitutions, offering a distant parallelism with the change (1) → (2) of p. 2. The extensive series of rearrangements studied by Smiles and by Truce are mainly of the following type:

$$o\text{-}(\text{Ph—X})\text{C}_6\text{H}_4\text{—Y}^- \longrightarrow o\text{-}(\text{X}^-)\text{C}_6\text{H}_4\text{—Y—Ph}$$

These reactions are understandably favoured if Y^- is more basic than X^-, and are also promoted by electron-attracting substituents in the right-hand benzene nucleus; an early, typical example is the rearrangement of (9).

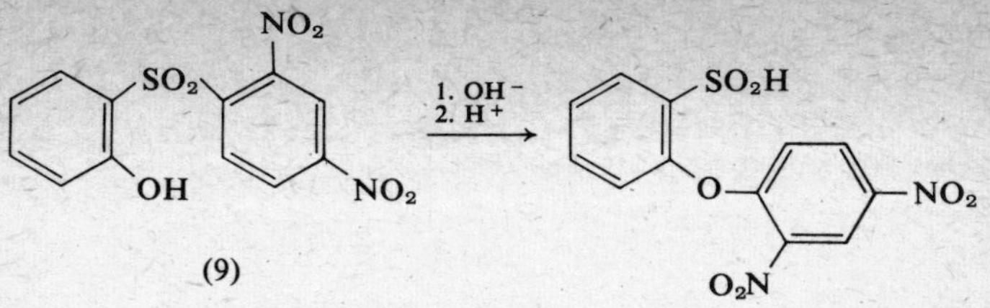

(9)

Numerous cases are known in which, under acid conditions, an atom or group migrates from a nitrogen atom to an attached benzene nucleus. Since these involve *o*- or *p*- attack on a benzene ring carrying a nitrogen substituent, they recall the phenomena of electrophilic substitution. Examples are: the Fischer–Hepp rearrangement of nitrosamines; the rearrangement of nitramines; the diazoamino- → aminoazo-benzene change; the Hofmann–Martius migration of alkyl groups from ammonium nitrogen

$$(\mathrm{Ph\overset{+}{N}Me_3\overset{-}{Cl}} \xrightarrow{300^\circ} \mathrm{MeC_6H_4NMe_2{\cdot}HCl});$$

the Orton rearrangement of *N*-haloanilides; the benzidine change.

In spite of their formal similarity, these reactions do not all 'belong together'. Thus, while the rearrangement of *N*-chloroacetanilide is *inter*molecular with molecular chlorine splitting off intermediately, the acid-promoted conversion of phenylnitramine, $PhNHNO_2$, into *o*-nitroaniline is at least predominantly *intra*molecular. This is also true of the still more enigmatic benzidine and associated changes.

'No-mechanism' Rearrangements

There exists a series of transformations which the baffled theorists of a decade ago described as 'no-mechanism' rearrangements, but which are now more hopefully designated *electrocyclic* and *sigmatropic* processes. Characteristically, they are little affected by polar influences and, at the same time, show no evidence of the intervention of free radicals.

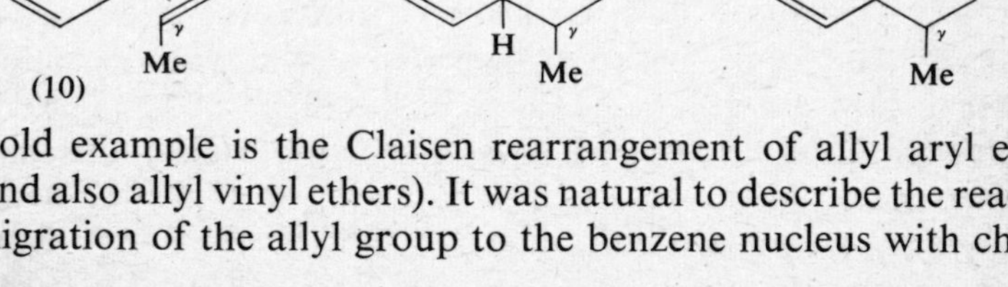

(10)

An old example is the Claisen rearrangement of allyl aryl ethers (10) (and also allyl vinyl ethers). It was natural to describe the reaction as a migration of the allyl group to the benzene nucleus with change of the point of attachment from C_α to C_γ. But it would also be logical, if apparently perverse, to assert that the phenoxyl group migrates from C_α to C_γ of the allyl system with change of its point of attachment from O to C_{ortho}!

One can draw a general picture

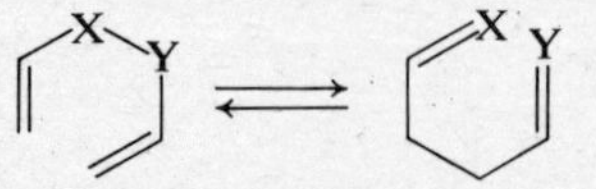

which is not in any way limited to ethers, as Cope has shown. It seems a natural extension to replace the oxygen atom of Claisen's ethers by a carbon atom bearing strongly electron-attracting groups, and the substance (11) does in fact rearrange as indicated.

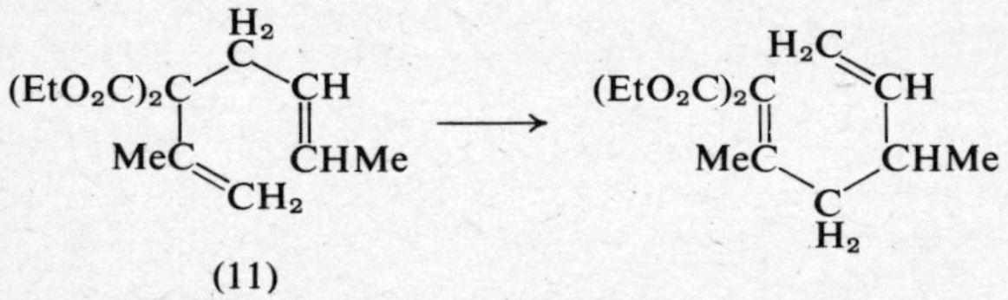

(11)

Further study showed that the electron-attracting ester groups, though promoting the reaction (increased conjugation in the product), are not essential. Perhaps the ultimate simplification of this very general Claisen/Cope series of rearrangements is the reversible process (12 ⇄ 13).

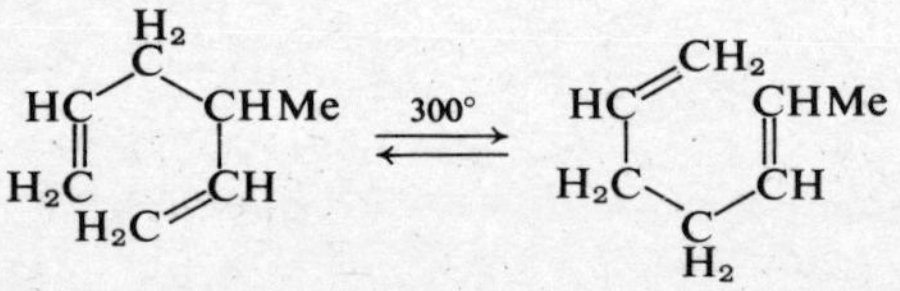

(12) (13)

These are examples of a wide range of changes which can be described, in old-fashioned terms, as proceeding by rearrangement of bonds, without migration of atoms or groups. They can be effected rather easily when they lead to relief of strain or increase in conjugation, especially in cyclic systems which impose a favourable conformation. They show a high degree of stereospecificity which is strikingly accommodated by the Woodward–Hoffmann rules. Alternatively, analogous reactions can be achieved photochemically with, in accordance with the rules, reversal of the stereochemical outcome. Thus, this type of reaction is of great value for the planned stereospecific syntheses of polycyclic systems.

Unambiguous classification is not always easy. The Chapman rearrangement of imino-ethers

$$\underset{\text{PhC=NAr'}}{\overset{\text{OAr}}{|}} \xrightarrow{280^\circ} \underset{\text{PhC—NArAr'}}{\overset{\text{O}}{\|}}$$

that proceeds with reluctance but strictly in an intramolecular sense, can be regarded plausibly either as a 1,3-electrophilic migration or as a [1, 3] sigmatropic process. Each view involves a 4-centre transition state, and the question arises: are the two interpretations irreconcilable—or indistinguishable?

Current Trends

Molecular rearrangements are still under active investigation, leading to a constantly developing picture. It will now be understood that they are not, in general, immediately predictable from *inter*molecular experience; each type is, at the outset, a phenomenon *sui generis* to be studied for its own sake. Indeed, even when two changes appear to be formally similar, they may proceed by very different mechanisms.

The use of more powerful physico-chemical methods is exemplified by the application of 'C.I.D.N.P.' (chemically induced dynamic nuclear polarization) to show that 1,2-electrophilic migration involves homolytic fission. In the Wittig rearrangement

$$\begin{matrix} Ph_2\bar{C}\text{—O} \\ \quad | \\ \quad CH_2Ph \end{matrix} \longrightarrow \left[\begin{matrix} Ph_2\bar{C}\text{—O}\cdot \\ \\ \cdot CH_2Ph \end{matrix} \right] \longrightarrow \begin{matrix} Ph_2C\text{—}\bar{O} \\ | \\ CH_2Ph \end{matrix}$$

the free-radical intermediates are held in a solvent cage before recombining in another sense; the postulated radical-anion is that of the well-known stable metal ketyl (Chapter 5).

Precise kinetic studies applied to materials substituted with deuterium at strategic points have thrown considerable light on mechanism. A rather sophisticated example of the use of stereochemical methods is offered by Doering's study of a Cope rearrangement (Chapter 8).

Application to Synthesis

Many rearrangements proceed smoothly and give good yields of product, so that they find valuable application in synthesis. It is significant that trimethylacetic acid owes its trivial name, pivalic acid (*pi*nacolone *val*er*ic* acid), to its accessibility by a sequence including a pinacolic rearrangement. Earlier came the exploitation of Hofmann's still valuable method for the conversion of acid amides into amines. As later examples can be cited Pictet's classical synthesis of nicotine (*N*-arylpyrrole → 2-arylpyrrole), Haworth's synthesis of cryptopine and protopine (transannular migration of oxygen), and electrocyclic processes leading with comparative ease to quite recherché fused polycyclic structures (see Chapter 7). The biogenetic occurrence of Wagner–Meerwein changes has long been recognised, and electrophilic rearrangements of 'onium ions have been postulated as biosynthetic routes to long-chain structures. The following chapters will include numerous examples of rearrangements which can constitute serviceable synthetic methods.

References

1. F. C. Whitmore, *J. Amer. Chem. Soc.*, 1932, **54**, 3274.
2. See R. K. Friedlina, *Adv. Free Radical Chem.*, 1965, **1**, 211.

Review Articles

Molecular Rearrangements, Parts 1 and 2, ed. de Mayo, Interscience (New York), 1963.

Mechanisms of Molecular Rearrangements, several volumes, ed. Thyagarajan, Interscience (New York), 1968, in progress.

2

Nucleophilic Rearrangements I: Migration from Carbon to Carbon

The migration of a group from one site to another elsewhere in a molecule is a rearrangement process commonly encountered in organic chemistry. The migration terminus in *nucleophilic* rearrangements is an atom which is either electron-deficient (e.g. C^+) or potentially so in the sense that it bears a group which can be expelled as an anion. These migrations are thus differentiated from *electrophilic* rearrangements, discussed in Chapters 5 and 6, in which migration of a group to an electron-rich terminus (e.g. C^-) occurs. These classes of molecular rearrangement are also sometimes referred to as anionotropic and cationotropic respectively. These descriptions are rather misleading, however, since reactions involving the movement of groups as free anions or cations are rare.

The most common type of nucleophilic rearrangement involves an intramolecular shift of a group from one atom to a neighbouring atom in the substrate. The operation of such a process was first properly recognised [1] by Whitmore and the rearrangement is sometimes described as a Whitmore 1,2-shift. Longer-range migrations are much less frequently encountered and usually occur only in situations where the molecular stereochemistry is particularly suited to a rearrangement of this type (e.g. in cyclic systems where transannular reactions are possible).

For a 1,2-nucleophilic migration, the sequence of events can be depicted by the following general scheme:

$$\underset{\substack{|\\X}}{\overset{\substack{R\\|}}{A}\text{—}B} \xrightarrow{-X} \overset{\substack{R\\|}}{A}\text{—}B \longrightarrow A\text{—}\overset{\substack{R\\|}}{B} \xrightarrow{+Y} A\text{—}\underset{\substack{|\\Y}}{\overset{\substack{R\\|}}{B}}$$

The process then involves several steps which, dependent upon the particular reaction under consideration, may or may not proceed independently, as we shall see later.

Step 1 involves the generation of an electron-deficient atom B, the migration terminus, usually through elimination of a leaving-group X which departs with its bonding electrons. The reactive site B possesses an incomplete valence shell of six electrons (an 'open sextet') and may be a positively charged carbon, nitrogen, or oxygen atom or an electrically neutral carbenoid or nitrenoid centre.

Step 2 involves the shift of the migrating-group R from atom A, the migration origin, to atom B. This step may follow or proceed simultaneously with Step 1. In the latter situation, of course, electron deficiency at the migration terminus is prevented by concerted migration of R with elimination of X. Such a synchronous process may be thought of as a nucleophilic substitution of X by R, i.e., an internal S_N2 reaction. The consecutive process, by a similar analogy, may be likened to an S_N1 substitution at B. In this and the following two chapters, the discussion will be almost completely restricted to reactions where R is a migrating carbon group, i.e., where there is a rearrangement of the carbon skeleton of the molecule. Thus, although many examples of 1,2-hydrogen shifts have been discovered, these will be considered only in relation to the mechanism of carbon migration. Related rearrangements in which the migrating-group is a hetero-atom (e.g., OH, NR_2, halogen) will not be discussed. [See J. March, *Advanced Organic Chemistry: Reactions, Mechanisms, and Structure*, McGraw-Hill, New York (1968), p. 813 *et seq.* for a discussion of reactions of this type.]

Step 3. Addition of a nucleophilic species Y to atom A often terminates the reaction. This process may occur simultaneously with Step 2 and, in special cases, Steps 1, 2, and 3 may all take place in a concerted fashion. As will become apparent, however, other reactions (e.g., proton elimination) may also occur at this stage.

In order to obtain an understanding of the factors controlling the course of nucleophilic rearrangements, several questions are posed that concern the mechanism of the reaction. These may be itemised as follows.

(a) What factors control the readiness with which a group migrates?

(b) What are the stereochemical and conformational requirements of the rearrangement?

(c) What are the stereochemical consequences of the rearrangement at the migrating-group, the migration origin, and the migration terminus?

(d) Does the actual migration (Step 2) involve formation of an intermediate or merely a transition state?

(e) Do any of the steps in the rearrangement sequence take place simultaneously?

Evidence relating to all of these questions has been obtained and the factors controlling 1,2-nucleophilic shifts are now well understood. In this chapter, migrations from carbon to carbon will be considered (i.e., carbonium-ion and related rearrangements). In the following two chapters, the related carbon-to-nitrogen and carbon-to-oxygen rearrangements are discussed. The text is subdivided in this way as a matter of convenience but the inter-relationship of the subject material contained in these three chapters should be investigated and appreciated.

Nucleophilic Migrations from Carbon to Carbon

Carbonium-ion rearrangements have fascinated organic chemists and occupied their attention for well over a century, although the postulate of electron-deficient carbon intermediates did not achieve respectability until the 1920s. The great volume of research endeavour which has been devoted to this topic has contributed in no small measure to our present understanding of the chemical reactivity of organic molecules. As illustrated by the following examples, a large variety of carbonium-ion rearrangements involve a 1,2-shift. For the reactions cited, the migrating carbon atom is indicated by an asterisk.

Neopentyl and related rearrangements. The rearrangement of neopentyl compounds to t-amyl compounds was one of the first examples studied of a 1,2-nucleophilic rearrangement. The driving force for the migration of the methyl group may be attributed to the gain in stability associated with the formation of a tertiary from a primary carbonium ion. The reaction is initiated by elimination of a leaving-group and is terminated either by capture of a nucleophile or by elimination of a proton [2, 3]. For example:

$$\mathrm{Me_2\overset{\overset{\displaystyle Me^*}{|}}{C}CH_2OH \xrightarrow{HBr} Me_2\overset{\overset{\displaystyle Br}{|}}{C}CH_2Me^*}$$

$$\mathrm{Me_2\overset{\overset{\displaystyle Me^*}{|}}{C}CH_2Br \xrightarrow[EtOH]{EtO^-} Me_2\overset{\overset{\displaystyle OEt}{|}}{C}CH_2Me^* + Me_2C{=}CHMe^*}$$

A very large number of related rearrangements incorporating 1,2-shifts of aryl and alkyl groups are known. For example:

$$\underset{\displaystyle Ph_2CCH_2OH}{\overset{Ph^*}{|}} \xrightarrow{HCO_2H} \underset{\displaystyle Ph_2CCH_2Ph^*}{\overset{OCHO}{|}} + Ph_2C{=}CHPh^*$$

The mechanism of aryl migration is discussed in a later section.

Pinacol and related rearrangements. In 1860, Fittig reported [4] that pinacol (1) is converted to pinacolone (2) on treatment with aqueous sulphuric acid.

$$\underset{(1)}{\underset{\displaystyle MeC(OH)C(OH)Me_2}{\overset{Me^*}{|}}} \xrightarrow{H_2SO_4} \underset{(2)}{\underset{\displaystyle MeCOCMe_2}{\overset{Me^*}{|}}}$$

It was soon discovered that glycols generally undergo a skeletal rearrangement of this type, and many examples of such 1,2-shifts of both alkyl and aryl groups have since been found. The term **pinacol rearrangement** is now used to describe all reactions of the type

$$-\overset{R^*}{\overset{|}{C}}(OH)\overset{|}{\underset{|}{C}}(OH) \xrightarrow{H^+} -\overset{O}{\overset{\|}{C}}-\overset{R^*}{\overset{|}{\underset{|}{C}}}-$$

This rearrangement is initiated through acid-promoted removal of a hydroxyl group from the diol giving a β-hydroxycarbonium ion. Since cations of this type can be generated in other ways, it is not surprising to find a number of other reactions which are closely related to the pinacol rearrangement.

(a) *Deamination of β-hydroxyamines:*

$$-\overset{R^*}{\overset{|}{C}}(OH)\overset{|}{\underset{|}{C}}(NH_2) \xrightarrow{HNO_2} -\overset{O}{\overset{\|}{C}}-\overset{R^*}{\overset{|}{\underset{|}{C}}}-$$

When applied to the ring expansion of cycloalkylmethylamines, this reaction is referred to as the **Tiffeneau–Demjanov rearrangement**. For example:

$$\text{(1-hydroxycyclopentyl)}\ CH_2NH_2 \xrightarrow{HNO_2} \text{cyclohexanone}\ (=O)$$

(b) *Hydrolysis of β-halohydrins:* Removal of halide anion from a β-halohydrin either under hydrolysis conditions or assisted by silver ion leads to pinacol-like rearrangement.

$$-\overset{R^*}{\underset{}{C}}(OH)\underset{|}{\overset{|}{C}}(X) \xrightarrow[\text{or } Ag^+]{H_2O} -\overset{O}{\overset{\|}{C}}-\overset{R^*}{\underset{|}{C}}-$$

(X = Cl, Br or I)

(c) *Acid-promoted epoxide ring-opening:* Treatment of epoxides with aqueous mineral acid often leads to the formation of a rearranged product.

$$R^*\!-\!C\overset{O}{\frown}C \xrightarrow{H^+} -\overset{O}{\overset{\|}{C}}-\overset{R^*}{\underset{|}{C}}-$$

(d) *Acyloin rearrangement:* The rearrangement of α-hydroxy-ketones or -aldehydes on treatment with strong acid leads to a product in which the oxygen functions are interchanged and a group has migrated.

$$-\overset{R^*}{\underset{HO}{C}}-\overset{O}{\overset{\|}{C}}- \underset{}{\overset{H^+}{\rightleftharpoons}} -\overset{R^*}{\underset{HO}{C}}-\overset{\overset{+}{O}H}{\overset{\|}{C}}- \longleftrightarrow -\overset{R^*}{\underset{HO}{C}}-\overset{OH}{\overset{|}{C^+}}-$$

$$\updownarrows$$

$$-\underset{O}{\underset{\|}{C}}-\overset{R^*}{\underset{OH}{C}}- \overset{-H^+}{\rightleftharpoons} -\underset{HO^+}{\underset{\|}{C}}-\overset{R^*}{\underset{OH}{C}}- \longleftrightarrow -\underset{HO}{\overset{+}{C}}-\overset{R^*}{\underset{OH}{C}}-$$

This rearrangement may take a different course in situations where acid-promoted removal of the hydroxyl group can produce a stabilised carbonium-ion centre, e.g. in the rearrangement of benzoin to diphenylacetic acid.

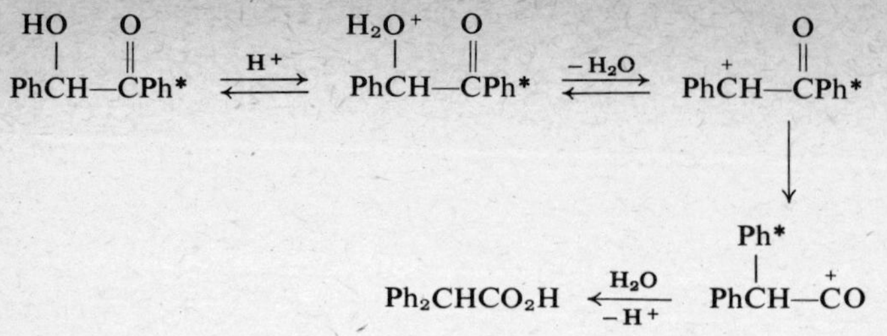

$$\underset{\text{PhCH—CPh*}}{\overset{\text{HO}\quad\text{O}}{}} \underset{}{\overset{H^+}{\rightleftharpoons}} \underset{\text{PhCH—CPh*}}{\overset{H_2O^+\quad\text{O}}{}} \underset{}{\overset{-H_2O}{\rightleftharpoons}} \overset{+}{\text{PhCH}}\text{—}\overset{\text{O}}{\text{CPh*}} \longrightarrow \text{PhCH(Ph*)—}\overset{+}{\text{CO}} \xrightarrow[-H^+]{H_2O} Ph_2CHCO_2H$$

Wagner–Meerwein rearrangement. In 1899, Wagner made the then revolutionary suggestion [5] that bornyl chloride (4), the major product of the reaction of α-pinene (3) with hydrochloric acid, possessed a rearranged carbon skeleton. Meerwein later demonstrated [6] that carbonium-ion intermediates were implicated in the camphene hydrochloride (5) → isobornyl chloride (6) rearrangement and pointed the way to an understanding of the rearrangements which had baffled earlier terpene chemists.

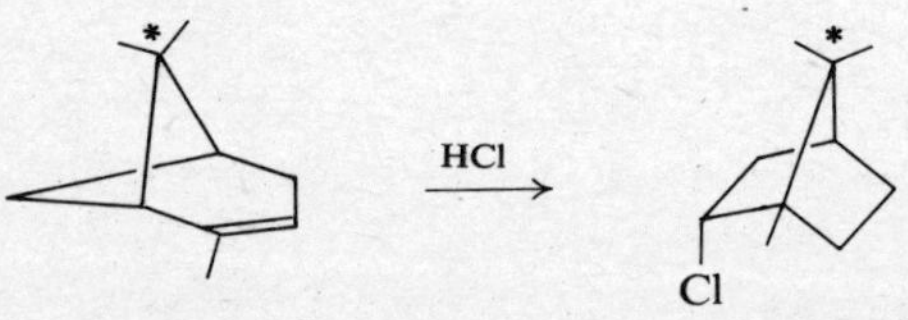

HCl

(3) (4)

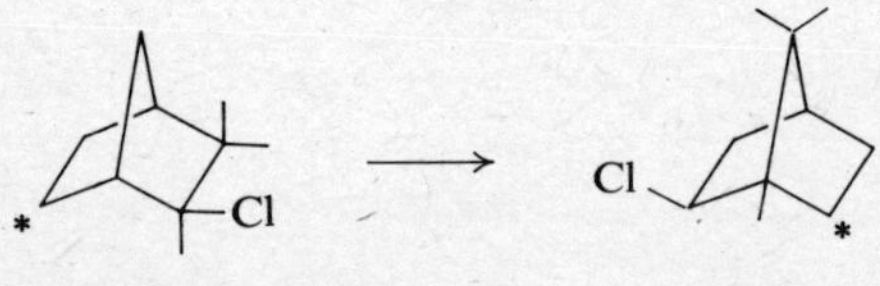

(5) (6)

The term Wagner–Meerwein rearrangement is now generally taken to describe a 1,2-migration of a ring carbon atom in a bridged polycyclic molecule. Some chemists have expanded the connotation to include *all* 1,2-migrations of alkyl or aryl groups from carbon to carbon.

Dienone–phenol rearrangement. Upon treatment with acid, 4,4-dialkylcyclohexa-2,5-dienones (7) rearrange to phenols (8) through migration of an alkyl group to an adjacent ring position. The driving force for this rearrangement may be attributed to the formation of an aromatic system.

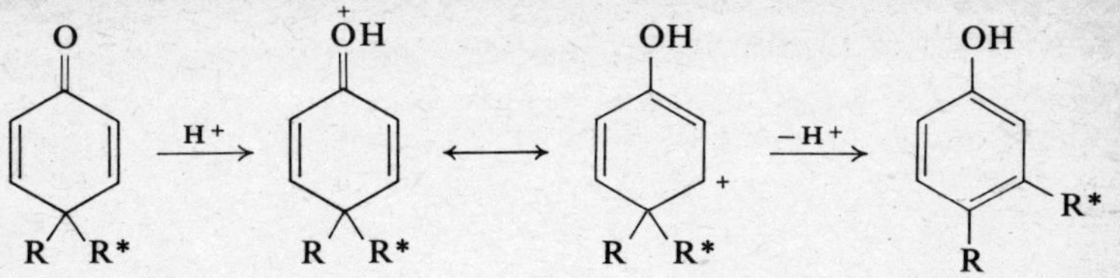

(7) (8)

Demjanov rearrangement. When a 1,2-migration is initiated through formation of a carbonium ion by diazotisation of a primary amine (i.e., $RNH_2 \rightarrow RN_2^+ \rightarrow R^+$), the reaction is termed a Demjanov rearrangement. The reaction has found particular application for the expansion of alicyclic rings by one carbon atom, although the Tiffeneau modification (*vide supra*) usually provides cleaner reactions. For example, treatment of cyclobutylmethylamine (9) with nitrous acid yields cyclopentanol (10) and cyclopentene (11) in addition to non-rearranged products [7].

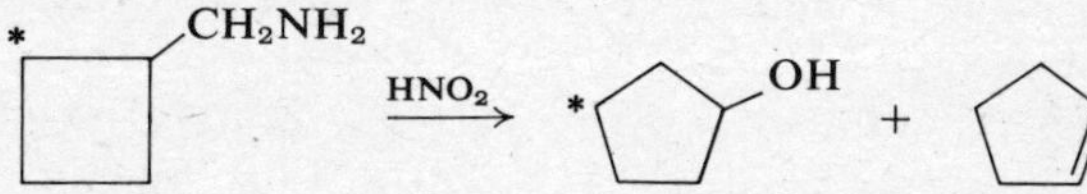

(9) (10) (11)

Nametkin rearrangement. As will become apparent later, 1,2-shift of a methyl group is a process commonly encountered in carbonium-ion rearrangements. The term Nametkin rearrangement has come to refer to such a rearrangement of bridged bicyclic monoterpenes and related compounds. For example, dehydration of the bicyclic alcohol (12) with potassium bisulphate leads to the formation of the rearranged alkene (13) [8].

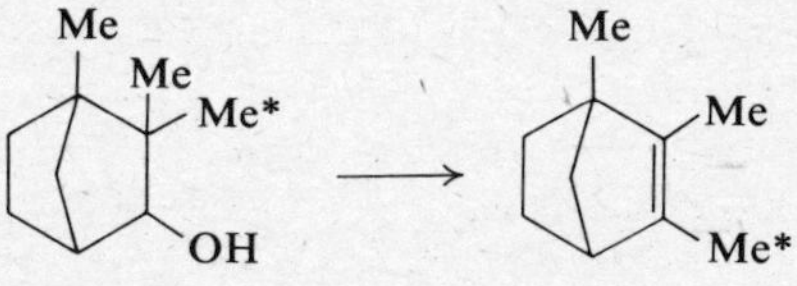

(12) (13)

Mechanistic Studies

The intramolecular nature of the vast majority of 1,2-nucleophilic migrations is well established. For example, treatment of a mixture of 1,2-diols with acid promotes pinacol rearrangement (*vide supra*). However, the only products formed in the reaction are those obtainable from rearrangement of the individual diols and no products of 'crossover' migration are formed.

$$\left.\begin{array}{c} R_2C(OH)C(OH)R_2 \\ + \\ R'_2C(OH)C(OH)R'_2 \end{array}\right\} \begin{array}{l} \xrightarrow{H^+} RCOCR_3 + R'COCR'_3 \\ \not\longrightarrow RCOCR_2R' \text{ or } R'COCR'_2R \end{array}$$

Consequently, the migrating-group never becomes completely detached from the substrate during the rearrangement.

Evidence relating to the intervention of carbonium-ion intermediates in the reactions so far considered appears overwhelming (see, however, pp. 25 and 31). The first clues were provided by the classical studies of Meerwein and his collaborators which were reported in a series of papers published from 1922 onwards. They showed, for example, that the rate of the rearrangement of camphene hydrochloride (5) to isobornyl chloride (6) was first-order and increased with the increase in the ionising power of the solvent in which the reaction was conducted, i.e., a rate increase was observed through the solvent series: petrol < benzene < anisole < acetonitrile < sulphur dioxide. Further, the rate of the process was dramatically accelerated in the presence of catalysts such as $FeCl_3$, $SnCl_4$, and $HgCl_2$ which were known to complex readily with chloride anion. In the light of these results, the conclusion that a cationic intermediate is formed during the reaction is inescapable.

A further argument supporting the concept of carbonium-ion intermediates was provided by Ingold, who pointed out that rearrangements that require the separation of an anion from a carbon atom (i.e., to generate a carbonium ion) should be facilitated by groups that can increase the electron supply to the reaction site. This proves to be so. With reference to the pinacol rearrangement, for example, acid-promoted elimination of a hydroxyl group from the diol (i.e. 14 →

$$\underset{(14)}{R_2C(OH)\text{—}CR_2(OH)} \underset{}{\overset{H^+}{\rightleftharpoons}} \underset{(15)}{R_2C(\overset{+}{O}H_2)\text{—}CR_2(OH)} \overset{-H_2O}{\rightleftharpoons} \underset{(16)}{R_2\overset{+}{C}\text{—}CR_2(OH)}$$

15 → 16) is assisted when electron-donating groups are attached to the ionisation site (e.g., 14; R = $p\text{-MeOC}_6H_4$).

Further evidence for carbonium-ion formation in the pinacol rearrangement has been obtained by oxygen-exchange experiments [9]. Partial rearrangement of pinacol (1) to pinacolone (2) was carried out in acidic solutions containing $H_2{}^{18}O$. It was found that the recovered pinacol (18) contained isotopically labelled oxygen, which established that formation of the carbonium ion (17) from pinacol is a reversible process as indicated in Scheme 1. Further, the relative proportions of diol and ketone present in solution were directly related to the acidity of the medium. The relative rates of rearrangement and oxygen exchange showed that, at low acidities, about 75 per cent of the generated carbonium ions reverted to diol and the remaining proportion rearranged to pinacolone.

Scheme 1

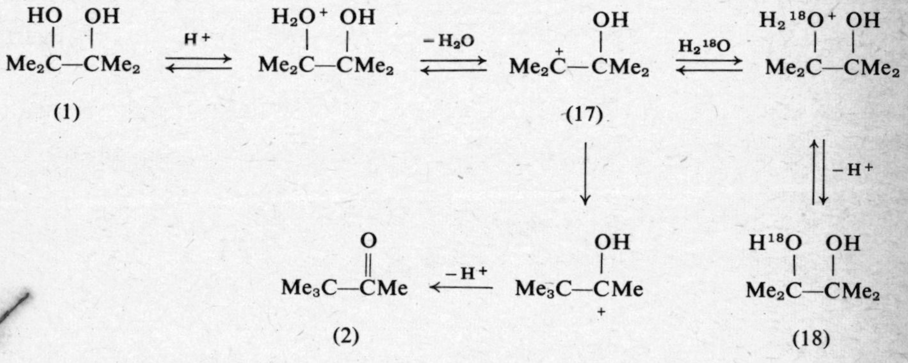

$$\underset{(1)}{\text{Me}_2\overset{\text{HO}}{\text{C}}\text{—}\overset{\text{OH}}{\text{C}}\text{Me}_2} \underset{}{\overset{\text{H}^+}{\rightleftarrows}} \text{Me}_2\overset{\text{H}_2\text{O}^+}{\text{C}}\text{—}\overset{\text{OH}}{\text{C}}\text{Me}_2 \overset{-\text{H}_2\text{O}}{\rightleftarrows} \underset{(17)}{\text{Me}_2\overset{+}{\text{C}}\text{—}\overset{\text{OH}}{\text{C}}\text{Me}_2} \overset{\text{H}_2{}^{18}\text{O}}{\rightleftarrows} \text{Me}_2\overset{\text{H}_2{}^{18}\text{O}^+}{\text{C}}\text{—}\overset{\text{OH}}{\text{C}}\text{Me}_2$$

$$(17) \downarrow \qquad\qquad \text{Me}_2\overset{\text{H}_2{}^{18}\text{O}^+}{\text{C}}\text{—}\overset{\text{OH}}{\text{C}}\text{Me}_2 \ \overset{-\text{H}^+}{\rightleftarrows}\ (18)$$

$$\underset{(2)}{\text{Me}_3\text{C—}\overset{\text{O}}{\overset{\|}{\text{C}}}\text{Me}} \overset{-\text{H}^+}{\longleftarrow} \text{Me}_3\text{C—}\underset{+}{\overset{\text{OH}}{\text{C}}}\text{Me} \qquad \underset{(18)}{\text{Me}_2\overset{\text{H}^{18}\text{O}}{\text{C}}\text{—}\overset{\text{OH}}{\text{C}}\text{Me}_2}$$

As mentioned previously, cations of the type (16) can be generated from precursors other than 1,2-diols. In this connection, it has been found that under similar experimental conditions, mixtures of pinacol (1) and pinacolone (2) *in the same relative proportions* are formed from different precursors, as shown in Scheme 2. Although the rates of these reactions are widely different, a common carbonium-ion intermediate (17) is clearly involved.

In many of the reactions so far considered, the carbonium-ion site is predetermined by the nature of the molecule. In the pinacol rearrangement of unsymmetrically substituted ethane-1,2-diols or ethylene oxides, in principle two different carbonium ions can be formed. It is not surprising to find, however, that such reactions proceed via the more stable of the two cations, provided, of course, that there is

Scheme 2

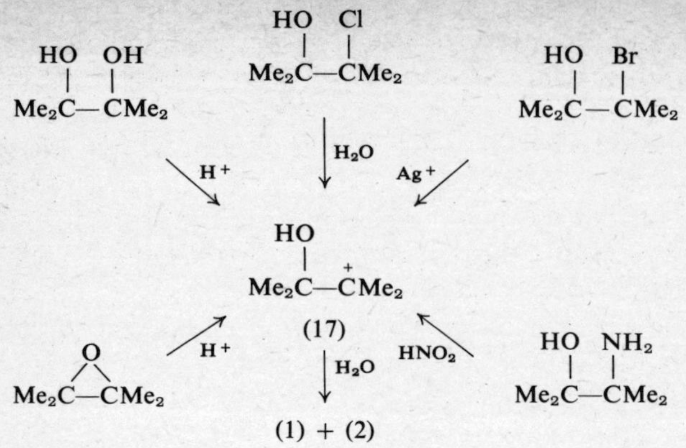

$$\mathrm{Me_2C(OH)\!-\!CMe_2(OH)} \xrightarrow{H^+} (17) \quad \mathrm{Me_2C(OH)\!-\!CMe_2Cl} \xrightarrow{H_2O} (17) \quad \mathrm{Me_2C(OH)\!-\!CMe_2Br} \xrightarrow{Ag^+} (17)$$

$$\text{Me}_2\text{C(O)CMe}_2\ (\text{epoxide}) \xrightarrow{H^+} (17) \quad \mathrm{Me_2C(OH)\!-\!CMe_2NH_2} \xrightarrow{HNO_2} (17)$$

$$(17)\ \mathrm{Me_2C(OH)\!-\!\overset{+}{C}Me_2} \xrightarrow{H_2O} (1) + (2)$$

an appreciable energy difference between the transition states preceding the formation of these species. It will be recalled that arylcarbonium ions $\left(\mathrm{Ar\overset{+}{C}{<}}\right)$ are more readily generated than alkylcarbonium ions $\left(\mathrm{R\overset{+}{C}{<}}\right)$ and that, for a series of related carbonium ions, stability increases through the series: primary $(\mathrm{-\overset{+}{C}H_2})$ < secondary $(\mathrm{-\overset{+}{C}H-})$ < tertiary $\left(\mathrm{-\overset{+}{C}{<}}\right)$. Bearing these factors in mind, the course of the following pinacol rearrangements can be readily understood.

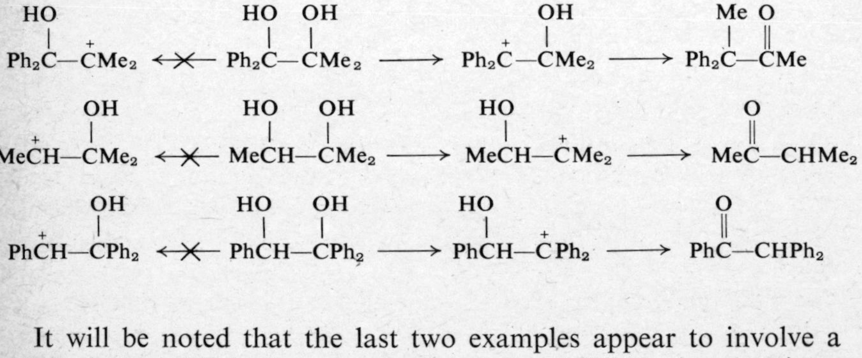

$$\mathrm{Ph_2C(OH)\!-\!\overset{+}{C}Me_2} \;\not\leftarrow\; \mathrm{Ph_2C(OH)\!-\!CMe_2(OH)} \longrightarrow \mathrm{Ph_2\overset{+}{C}\!-\!CMe_2(OH)} \longrightarrow \mathrm{Ph_2C(Me)\!-\!C(=O)Me}$$

$$\mathrm{Me\overset{+}{C}H\!-\!CMe_2(OH)} \;\not\leftarrow\; \mathrm{MeCH(OH)\!-\!CMe_2(OH)} \longrightarrow \mathrm{MeCH(OH)\!-\!\overset{+}{C}Me_2} \longrightarrow \mathrm{MeC(=O)\!-\!CHMe_2}$$

$$\mathrm{Ph\overset{+}{C}H\!-\!CPh_2(OH)} \;\not\leftarrow\; \mathrm{PhCH(OH)\!-\!CPh_2(OH)} \longrightarrow \mathrm{PhCH(OH)\!-\!\overset{+}{C}Ph_2} \longrightarrow \mathrm{PhC(=O)\!-\!CHPh_2}$$

It will be noted that the last two examples appear to involve a 1,2-hydrogen shift in the rearrangement step. While hydrogen migrations have been unambiguously established in many instances and may, in fact, occur in these pinacol rearrangements, an alternative mechanism requires consideration. By a process which has been termed 'vinyl dehydration', elimination of a proton from the intermediate cation leads directly to the enol tautomer of the product; that is

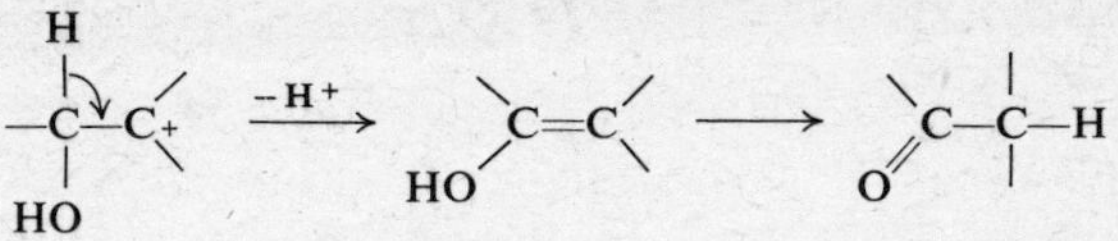

$$\mathrm{H{-}\overset{|}{C}(HO){-}C^{+}\ \xrightarrow{-H^{+}}\ (HO)C{=}C\ \longrightarrow\ O{=}C{-}\overset{|}{C}{-}H}$$

Migratory Aptitudes

All groups do not migrate with equal ease in 1,2-nucleophilic rearrangements. Following the sequence of events during the migration, at a certain stage (19) the migrating-group becomes partially bonded both to the migration origin and to the migration terminus, as indicated.

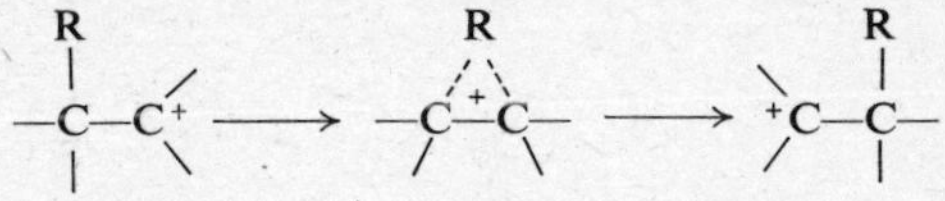

$$\mathrm{R{-}C{-}C^{+}\ \longrightarrow\ C\overset{R}{\cdots}C^{+}\ \longrightarrow\ {}^{+}C{-}C{-}R}$$

(19)

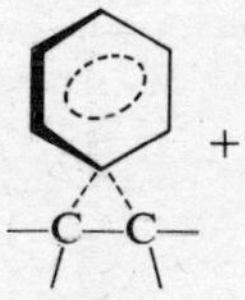

(20)

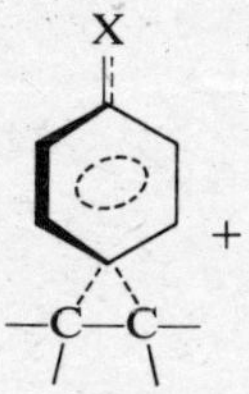

(21)

Dependent upon the nature of the group R, the structure (19) may represent either a transition state or an actual intermediate in the reaction. We shall return to this situation later. Irrespective of this point, however, it is obvious that the greater the ability of the group R to disperse the positive charge associated with the species, the lower will be the free-energy barrier for the migration. In this respect, aryl groups are much more effective than alkyl groups since the positive charge can be delocalised by the π-electron system of the benzene ring (cf. 20). In consequence, aryl groups migrate in preference to alkyl groups. For example, $MeCOCPh_2Me$ is the sole product of pinacol rearrangement of PhMeC(OH)C(OH)PhMe.

In examples where migration of two different aryl groups may occur, competition between the different rearrangement pathways is usually more finely balanced and mixtures of isomeric products are formed. This observation led to the idea that a group could be assigned a *migratory aptitude* that would represent the relative ease with which that group migrated. Bachmann and his collaborators, in particular, followed this approach [10] and estimated the relative proportions of the isomeric ketones (23 and 24) formed upon pinacol rearrangement of tetraarylethane-1,2-diols of the type (22). This product ratio was taken as the ratio of the migratory aptitudes of the groups in question, the phenyl group being assigned a reference value of 1·0. Migratory aptitudes determined in this manner were found to be remarkably consistent, irrespective of the diol used, and a number of examples are given in the Table.

$$\underset{(22)}{\mathrm{Ar{-}C(OH)(Ar'){-}C(OH)(Ar'){-}Ar}} \xrightarrow{H^+} \underset{(23)}{\mathrm{ArCOCAr'_2Ar}} + \underset{(24)}{\mathrm{Ar'COCAr_2Ar'}}$$

Relative Migratory Aptitudes of Groups in Pinacol Rearrangements

Group	Migratory aptitude
C_6H_5	1·0
p-$MeOC_6H_4$	500
p-MeC_6H_4	15·7
p-PhC_6H_4	12
m-$MeOC_6H_4$	1·6
p-ClC_6H_4	0·7
o-$MeOC_6H_4$	0·3
m-ClC_6H_4	0·1

The diols (22) used by Bachmann in this painstaking study contain two asymmetrically substituted carbon atoms and consequently can be obtained in *meso* and racemic modifications. Unfortunately, mixtures of these stereoisomers in undetermined proportions were employed and, as we shall see later, the course of 1,2-rearrangements is extremely sensitive to stereochemical influences. However, with this qualification, the trends exhibited by the values in the Table are readily understood. Electron-donating aryl substituents (e.g. *p*-MeO, *p*-Ph, *p*-Me) permit further delocalisation of positive charge in the rearrangement intermediate (21) and accordingly raise the migratory aptitude of the group relative to phenyl. Electron-withdrawing substituents (e.g. *m*- and *p*-Cl) disfavour delocalisation of positive charge to the benzene ring and exert the opposite effect. The relatively low migratory aptitude found in the case of the *o*-anisyl group probably reflects steric interference between the methoxyl substituent and the other aryl groups present. This interaction may cause the *o*-anisyl ring to twist from the conformation in which participation by the ring π-system is maximised (see 20), and the migratory aptitude of the group is correspondingly diminished.

Similar results have been found concerning the relative ease of migration of alkyl groups in the pinacol rearrangement [11]. An ethyl group migrates 17 times more readily and a t-butyl group > 4000 times more readily than a methyl group, for example. Electron release (inductive and/or hyperconjugative) from the methyl groups attached to the migrating carbon atom obviously provides assistance to the rearrangement, although on formation of the transition state in the migration the relief of steric chain may be a contributory factor, particularly in the t-butyl case.

Finally, in this connection, it must be emphasised that the migratory aptitudes given in the Table refer only to the pinacol rearrangement. While similar patterns of reactivity are found in other 1,2-rearrangements, the relative migratory aptitudes of groups are influenced by the nature of the reaction.

Stereochemical and Conformational Factors

We have seen that 1,2-nucleophilic rearrangements are initiated by expulsion of a leaving-group from the substrate and, in many reactions, this process precedes the actual migration step. In other cases, however, it is clear that the migrating-group R must begin to shift towards the migration terminus before the leaving-group X has completely departed. In other words, the reaction can be considered as an internal nucleophilic displacement of X by R. It is not surprising, therefore, that the most favourable stereochemistry for such a concerted elimination/migration process is that in which the migrating-group and the leaving-group are aligned in an *anti*-periplanar arrangement, as shown below.

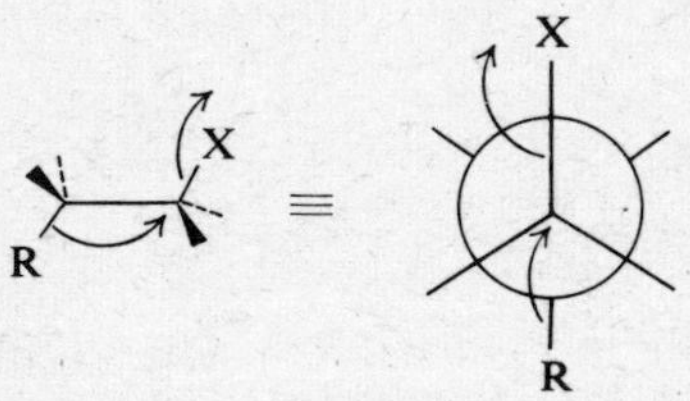

X ≡ X

R R

A clear-cut example of such stereoelectronic control of the rearrangement pathway is provided by the behaviour of the isomeric 2-amino-5-t-butylcyclohexanols (25–28) towards nitrous acid [12a]. Deamination of both the *cis*- (25) and the *trans*-hydroxyamine (26) is accompanied by ring contraction and the aldehyde (29) is formed exclusively in each case. Under the same reaction conditions, the other *cis*-hydroxyamine (27) is converted almost exclusively to the cyclohexanone (30) and the fourth compound of the series, the *trans*-isomer (28), gives yet another product, the epoxide (31). Since different products are formed from (25) and (28), these reactions cannot proceed via a free symmetrically solvated carbonium ion since the same cation (32) would be formed from either starting material. Similarly, deamination of the hydroxyamines (26) and (27) cannot involve production of the corresponding carbonium ion (33).

The presence of the bulky t-butyl substituent serves to hold the cyclohexane rings of these compounds in a rigid (non-inverting) chair conformation with the t-butyl group attached by an equatorial bond. Accordingly, the relative orientation of the amino and hydroxyl substituents can be clearly defined for each compound. Consider the conformations (25A–28A; represented both in 'sawhorse' and in Newman projection) of the diazonium ions formed in these reactions. From the structure of the resultant products it can be seen that it is the group which is located *anti*-periplanar to the leaving-group (N_2)

NH_2 Bu OH (25) $\xrightarrow{HNO_2}$ Bu, N_2^+, OH $\equiv$ Bu, N_2^+, OH (25A)

$-H^+$ | $-N_2$

Bu, CHO (29)

$-N_2$ | $-H^+$

NH_2 Bu OH (26) $\xrightarrow{HNO_2}$ Bu, N_2^+, OH $\equiv$ HO, Bu, N_2^+ (26A)

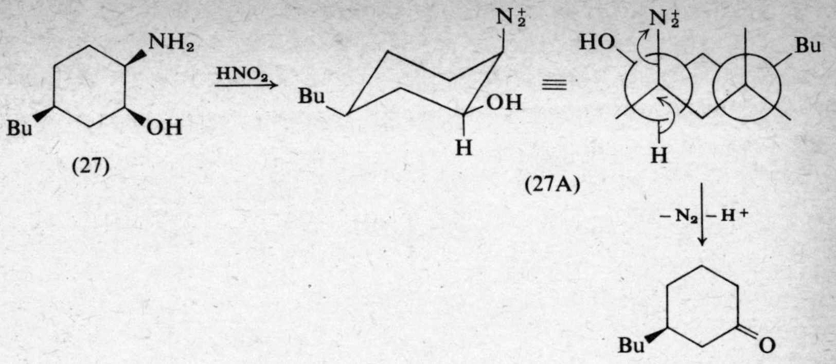

(27) (27A) (30) 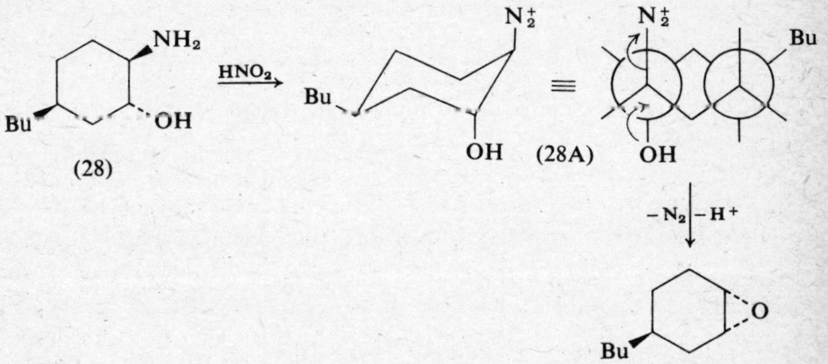 (28) (28A) (31) 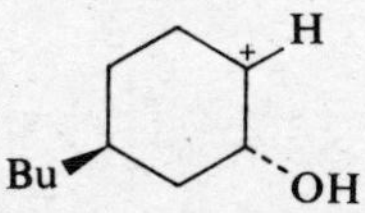 (32) 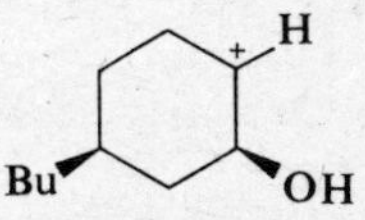 (33)

which is implicated in each individual rearrangement. Thus, for the diazonium ions (25A) and (26A), elimination of nitrogen must be accompanied by migration of the *anti*-periplanar ring carbon atom, leading to the same ring-contracted product (29) in each case. For the ions (27A) and (28A), however, ring contraction could only occur by migration of a carbon atom disposed *syn*-clinal to the leaving group, and this is not observed. The formation of the cyclohexanone (30) from the ion (27A) can be interpreted either in terms of a 1,2-hydrogen shift or by a 'vinyl dehydration' mechanism (see p. 23). In either case, it is the hydrogen atom located *anti*-periplanar to the leaving-group which is implicated. For the diazonium ion (28A), on the other hand, the *anti*-periplanar disposition of the hydroxyl group to the leaving-group leads predictably to the formation of the epoxide (31).

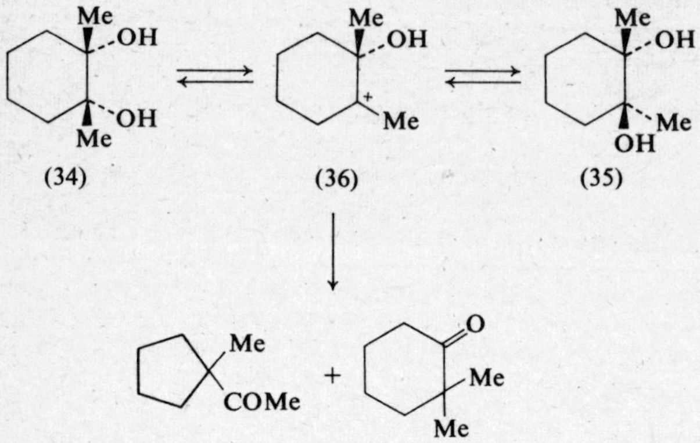

(34) (36) (35)

The results found in these deamination reactions contrast sharply with those found [12b] in the closely related pinacol rearrangements of the *cis*- and *trans*-cyclohexane-1,2-diols (34 and 35 respectively). Treatment of either diol with strong aqueous protic acid leads to the same product mixture consisting mainly of the ring-contracted product 1-acetyl-1-methylcyclopentane. Under these reaction conditions, acid-catalysed isomerisation of the diols was also established and it may be reasonably inferred that the rearrangement products derive from a common carbonium-ion intermediate (36). In these reactions, therefore, generation of this cation must precede the rearrangement step.

Many examples of stereospecific Whitmore 1,2-shifts have been encountered in the chemistry of bridged bicyclic compounds. A particularly striking example is provided by compounds of the nopinane series [13]. Thus, treatment of α-nopinol (37) with *p*-bromobenzenesulphonyl chloride in pyridine leads smoothly to the exclusive formation of *endo*-camphenilyl brosylate (39) and, under the same conditions, β-nopinol (40) is converted quantitatively to apobornyl brosylate (42). In these reactions, the first-formed nopinyl brosylates (38 and 41) must undergo ionisation followed by internal return of brosylate anion to a rearranged carbon framework. The driving force for these Wagner–Meerwein rearrangements is obviously the relief of strain obtained from disruption of the cyclobutane ring of the nopinyl system, and the structures of the products establish the exclusive migration of the carbon atom situated *anti* to the brosylate leaving group.

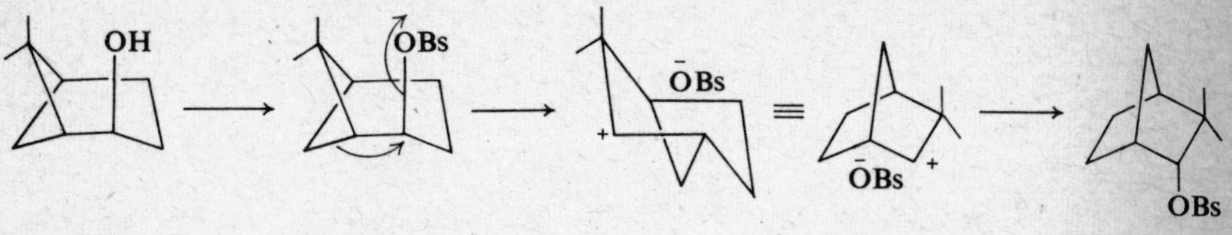

OH (37) → OBs (38) → $\bar{O}$Bs + ≡ $\bar{O}$Bs + → OBs (39)

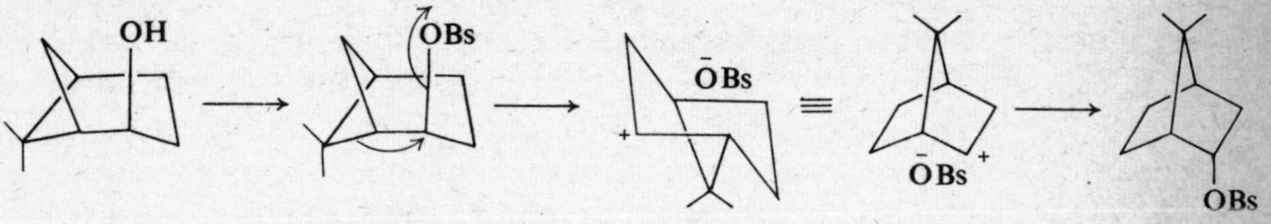

OH (40) → OBs (41) → $\bar{O}$Bs + ≡ $\bar{O}$Bs + → OBs (42)

Finally in this connection, remarkable skeletal rearrangements which involve a series of 1,2-shifts have been found in the chemistry of pentacyclic triterpenes. For example, on treatment with acid the hydrocarbon (43) of the friedelin series undergoes isomerisation to the hydrocarbon (44) which possesses a β-amyrin skeleton. The rearrangement [14] is initiated by protonation of the double bond in (43), then proceeds by a series of methyl (Nametkin) and hydrogen shifts, and is terminated by proton elimination as shown (45). Each migration occurs by *trans*-diaxial shift of migrating- and leaving-groups. Reactions of this type have come to be referred to as '**back-bone**' **rearrangements.**

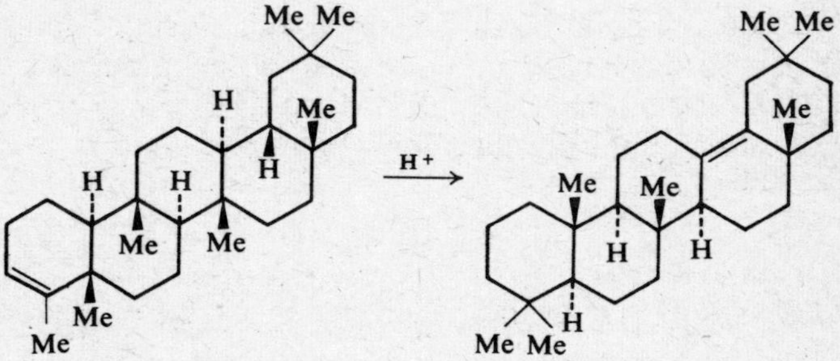

(43) $\xrightarrow{H^+}$ (44)

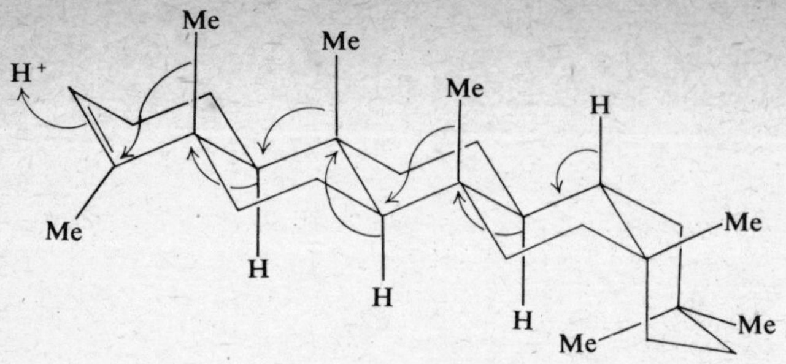

(45)

In the reactions thus far discussed, rotation is restricted around the C—C bond connecting the migration origin and migration terminus by their location in a cyclic structure. In acyclic systems, the substrate may adopt a number of conformations in which *trans*-1,2-rearrangements can take place. The particular conformation selected, in general, will be that in which unfavourable gauche steric interactions are minimised. This requirement sometimes leads to unexpected results. For example, the diastereoisomeric forms of the β-hydroxyamine (46) give different rearrangement products upon nitrous acid deamination [15]. Although migration of a *p*-anisyl group would have been expected to occur in each case (cf. migratory aptitude of *p*-anisyl relative to phenyl; see Table on p. 24), the diastereoisomer (47) rearranges by shift of a phenyl group to give the ketone (49). If we accept that the migration occurs when the departing nitrogen molecule is aligned *anti*-periplanar to the migrating group, shift of the *p*-anisyl group can take place only when the diazonium ion substrate adopts the conformation (48A). However, this conformation is destabilized by steric repulsion between the methyl group and the bulky *syn*-clinal aryl groups. Rotation around the C—C bond, however, permits the ion to adopt a more favourable conformation (48B) in which these interactions are relieved, and the rearrangement accordingly proceeds by migration of a phenyl group. Using a similar argument, the conformation (50B) of diastereoisomeric diazonium ion is preferred over the alternative conformation (50A) and the rearrangement takes place by shift of the *p*-anisyl group to give the ketone (51).

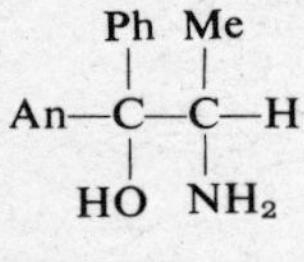

(46)

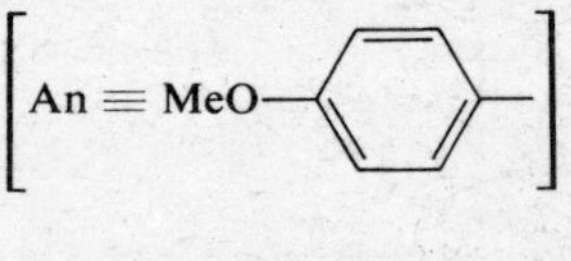

[An ≡ MeO—C_6H_4—]

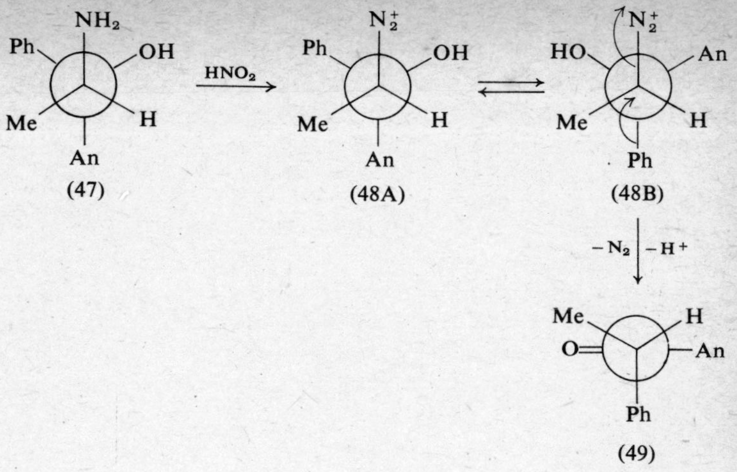

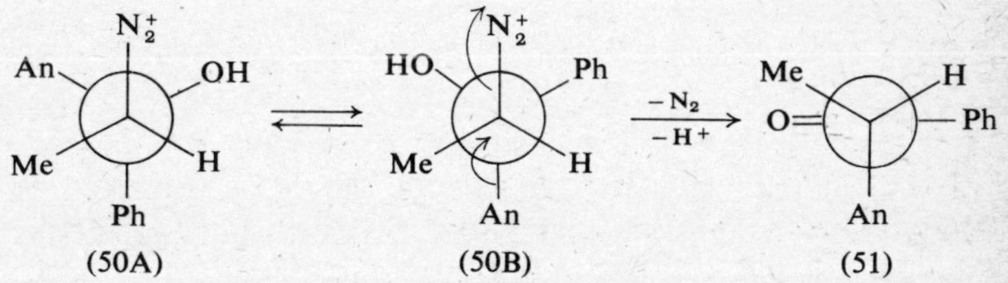

Timing of the Rearrangement Steps

The stereospecificity associated with the 1,2-rearrangements so far considered requires that departure of the leaving-group and shift of the migrating-group must occur simultaneously or almost simultaneously. If a fully developed carbonium ion were to be formed in these reactions, migration of a group to this electron-deficient centre must occur before conformational rearrangement of the substrate can take place (e.g., by rotation around the bond between the migration origin and migration terminus). However, if the carbonium-ion intermediate is sufficiently long-lived, rotational rearrangement of the ion can occur before migration takes place, and the stereochemical integrity of the rearrangement would obviously be lost. We shall see that both of these mechanisms do operate in some carbonium-ion rearrangements.

It should also be appreciated, however, that the processes described represent the extremes of a spectrum of reaction mechanisms. The extent of neighbouring-group participation by the migrating-group as the leaving-group departs (i.e., anchimerically assisted ionisation) may vary in importance, depending upon the particular reaction under consideration.

Scheme 3

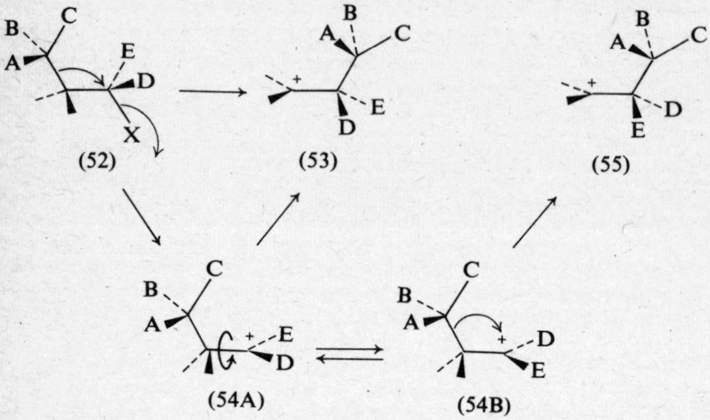

As illustrated in Scheme 3, the operation of a concerted or non-concerted (stepwise) rearrangement has important stereochemical consequences at the migration terminus. If a concerted (or near-concerted) rearrangement occurs (i.e., 52 → 53 or 54A → 53), inversion of configuration at the migration terminus must follow. If a stepwise rearrangement permits rotational equilibration (i.e. 54A ⇄ 54B), complete or partial racemisation at the migration terminus results (i.e., 54A + 54B → 53 + 55). For either mechanism, the migrating carbon atom shifts with retention of configuration.

Evidence relating to the retention of configuration at the migrating carbon atom has been obtained from studies of related carbon-to-nitrogen nucleophilic rearrangements, which are discussed in the following chapter. A demonstration that rearrangement is accompanied by inversion of configuration at the migration terminus has been provided [16] by a study of the rearrangement of the optically active β-hydroxyamine (56). Nitrous acid deamination of this compound yielded a ketonic product (57) which consisted predominantly of the enantiomer (57A) with inverted configuration at the migration terminus. The formation of minor amounts of the enantiomer (57B) with retained configuration at the migration terminus (the product was found to be about 25 per cent racemic), however, reveals that the fully developed carbonium ion (58) must also be formed in the reaction, and that migration of a phenyl group to each side of this electron-deficient centre can occur.

(56) $\xrightarrow{HNO_2}$ (diazonium ion) $\xrightarrow{-N_2}$ (58) ⇌ ; $\xrightarrow{-N_2,\ -H^+}$ (57A); (58) $\xrightarrow{-H^+}$ (57A); $\xrightarrow{-H^+}$ (57B)

Phenonium Ions

We have mentioned previously that, during a 1,2-nucleophilic rearrangement, the migrating-group becomes bonded both to the migration origin and migration terminus (see 19), and that such a structure may represent either a transition state or an actual intermediate bridged cation. A great deal of controversy surrounds the postulate of bridged carbonium-ion intermediates, particularly alkyl-bridged structures. [The alternative viewpoints are discussed in the following articles: H. C. Brown, *Chem. Brit.*, 1966, 199; G. D. Sargent, *Quart. Rev.*, 1966, **20**, 301; P. D. Bartlett, *Non-classical Ions*, Interscience, New York (1965); J. A. Berson in *Molecular Rearrangements*, Part 1, ed. de Mayo, Interscience, New York (1963).]

Phenyl-bridged carbonium ions are termed phenonium ions and we shall examine the evidence for their existence. Cram found [17] that the solvolysis of the optically active *threo*-tosylate (59) in acetic acid gave almost exclusively (96 per cent) the *threo*-acetate (60) that was found to be optically inactive. This result is nicely accommodated if it is considered that ionisation of the precursor (59) occurs with formation of the bridged cation (61). This intermediate possesses a plane of symmetry (i.e., the plane of the six-membered ring which is perpendicular to the plane of the three-membered ring) and thus the optical activity associated with the starting material is lost. Furthermore, capture of acetic acid by this intermediate can occur with equal ease either at the α- or the β-carbon atom, but only from the side opposite to the six-membered ring, to yield a racemic product (equal amounts of 60A and 60B). The formation of optically active *erythro*-acetate (96 per cent retention of configuration) in the acetolysis of the corresponding *erythro*-tosylate can be accommodated by a similar mechanism.

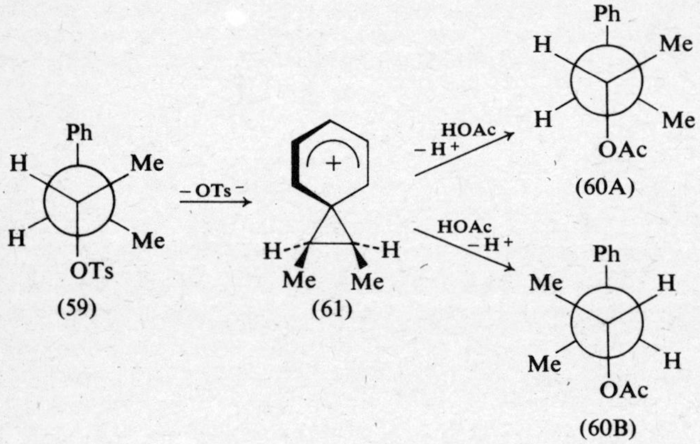

The most compelling evidence [18] for the existence of bridged phenonium ions has been obtained for compounds of the anthracene series. For example, hydrolysis of the tosylate (62) gave a mixture of alcohols (63 and 64) in which the rearranged product (63) predominated (85 per cent). This result is strongly suggestive of the presence of the bridged cation (65) which, in fact, has been characterised as the $(SbF_5OH)^-$ salt in strongly acidic media. The ^1H nuclear magnetic resonance spectrum of this salt showed clearly that the protons of the two methylene groups are equivalent. Furthermore, the chemical shift values observed for these protons indicated that little positive charge is borne by the methylene carbon atoms.

Kinetic results have also been interpreted in favour of phenyl-bridged cationic intermediates in solvolysis reactions. For example, 2,2,2-triphenylethylchloride (Ph_3CCH_2Cl) undergoes ionisation about

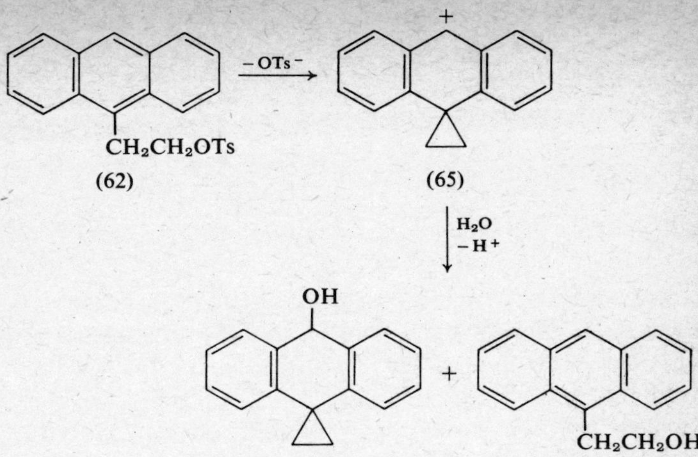

CH_2CH_2OTs (62) $\xrightarrow{-OTs^-}$ (65) $\xrightarrow[-H^+]{H_2O}$ OH (63) + CH_2CH_2OH (64)

10,000 times faster than 2,2-dimethylpropyl chloride (Me_3CCH_2Cl), and gives products of a rearranged carbon skeleton [19]. This enormous rate acceleration may be attributed to neighbouring-group participation by the phenyl group, which provides assistance to ionisation (see 66). However, the rate of this process may also be accelerated by relief of steric strain accompanying ionisation.

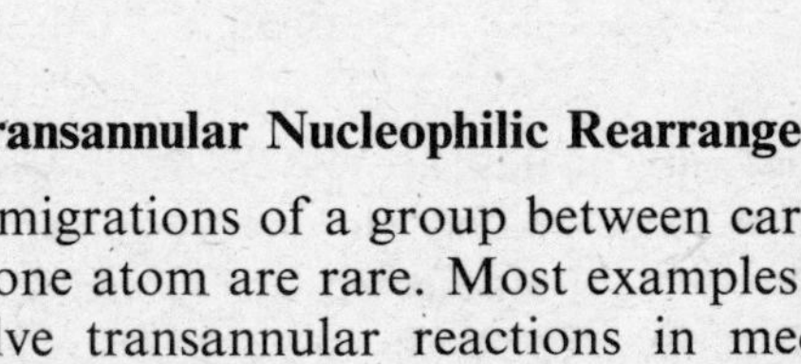

Ph Ph H H Cl $\xrightarrow[?]{-Cl^-}$ + Ph Ph H H

(66)

Transannular Nucleophilic Rearrangements

Longer-range migrations of a group between carbon atoms separated by at least one atom are rare. Most examples which have been discovered involve transannular reactions in medium-sized rings, usually by way of a hydride (or deuteride) shift (see A. C. Cope *et al.*, *Quart. Rev.*, 1966, **20**, 119). For example, upon treatment with phosphoric acid the cyclodecane-1,6-diol (67) is converted to the ketone (68). The mechanism of this reaction must involve formation of the cation (69) followed by transannular migration of deuteride as indicated.

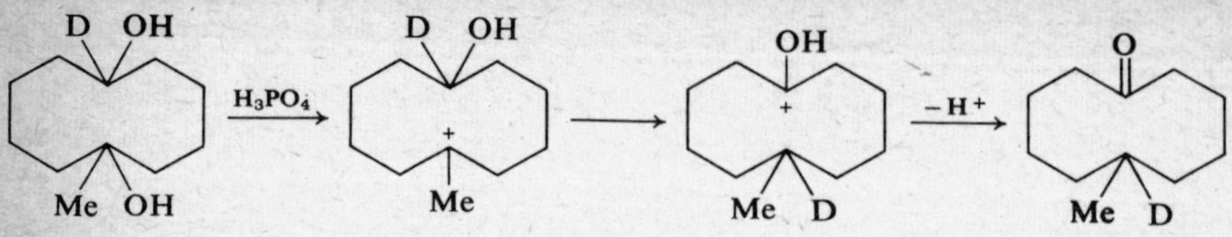

D OH
$\xrightarrow{H_3PO_4}$
Me OH
(67)
D OH
Me
(69)
OH
Me D
$\xrightarrow{-H^+}$
O
Me D
(68)

Transannular migration of an alkyl or aryl group clearly requires a much higher activation energy than that associated with a hydride shift, and very few examples of this type of reaction have been found. One such example was discovered [20] by Cope *et al.*, who found that the complex mixture of alkenes formed upon treatment of 5,5-diphenylcyclooctyl tosylate (70) contained about 1 per cent of the product of phenyl migration (71). The other products formed in this reaction were derived from the intermediate cation (72) either directly or following hydride shifts.

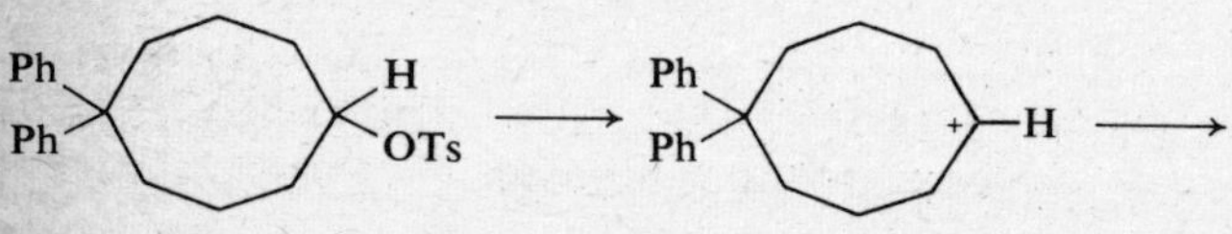

Ph Ph H OTs
(70)
Ph Ph H
(72)

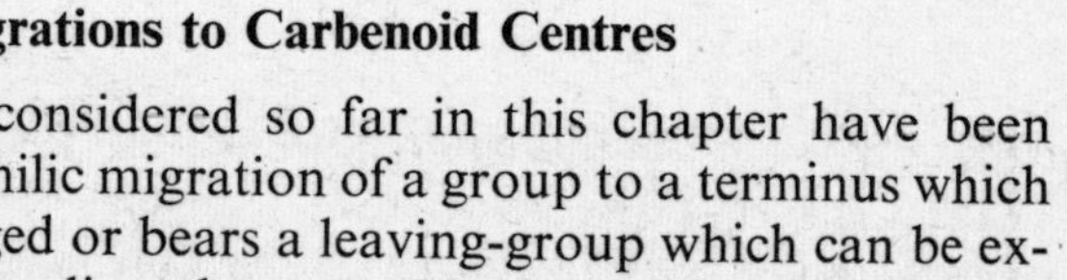

Ph Ph H
Ph Ph H
(71)

Migrations to Carbenoid Centres

The rearrangements considered so far in this chapter have been concerned with nucleophilic migration of a group to a terminus which is either positively charged or bears a leaving-group which can be expelled with its pair of bonding electrons. Nucleophilic migrations to electrically neutral carbenoid atoms are also known and, as we shall see, such rearrangements are particularly useful in organic synthesis. A reaction of this type was discovered, albeit unwittingly, by Wolff in 1902 although the true nature of the rearrangement was not clarified until much later [21]. In its original form, the reaction involved decomposition of an α-diazo-carbonyl compound ($RCOCR'N_2$) in hot water, whereupon a rearranged carboxylic acid ($RR'CHCO_2H$) was generated. It was later discovered that the reaction is greatly facilitated by the presence of silver compounds (e.g. Ag_2O), and that derivatives of the rearranged acid are formed when the rearrangement is conducted in protic non-aqueous media, e.g. esters or amides result when the reaction is carried in alcohols or amines respectively.

This reaction, now referred to as the **Wolff rearrangement**, is related in a mechanistic sense to the Curtius rearrangement (see next chapter) in which a carbon-to-nitrogen migration occurs. The application of the Wolff rearrangement to organic synthesis was developed by Arndt and Eistert who discovered [22] that the required α-diazo-ketones could be simply prepared by the reaction of diazoalkanes with carboxylic acid chlorides; that is:

$$RCOCl + 2R'CHN_2 \longrightarrow RCOCR'N_2 + R'CH_2Cl + N_2$$

The **Arndt–Eistert procedure** for the chain-lengthening of carboxylic acids may thus be generalised as shown in the following reaction sequence, and an example of the use of the method in synthesis is provided by the transformation 73 → 74 [23].

$$RCO_2H \longrightarrow RCOCl \longrightarrow RCOCR'N_2 \longrightarrow RR'CHCO_2H$$

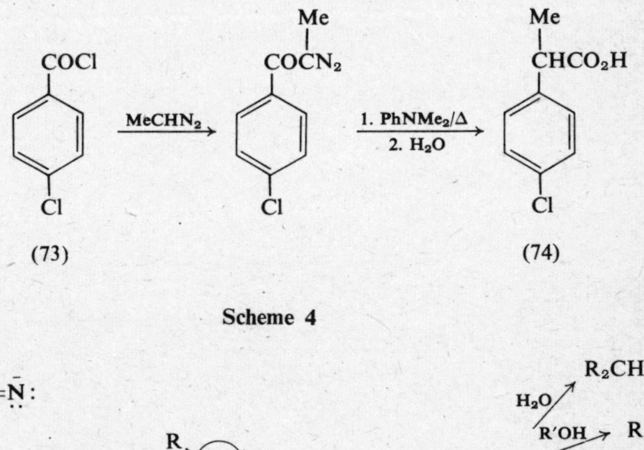

COCl, Cl (73) —$MeCHN_2$→ Me, COCN₂, Cl —1. $PhNMe_2/\Delta$; 2. H_2O→ Me, $CHCO_2H$, Cl (74)

Scheme 4

R, R, C—C=N⁺=N⁻:, O ↕ R, R, C—C⁻—N⁺≡N:, O (75) —Δ or hν, $-N_2$→ R, C—C—R, O (76) ⟶ $R_2C{=}C{=}O$ (77)

H_2O → R_2CHCO_2H

$R'OH$ → R_2CHCO_2R'

$R'NH_2$ → $R_2CHCONHR'$

NH_3 → $R_2CHCONH_2$

As mentioned previously, the catalytic activity of silver salts in the Wolff rearrangement is well established and we shall return to the mechanism of this process later. Under thermal or photochemical initiation in the absence of catalyst, the mechanism of the rearrangement of α-diazo-ketones and related compounds can be described as shown in Scheme 4. The formation of a carbenoid intermediate (76) by elimination of nitrogen from the diazo precursor (75) can be inferred from the observation that non-rearranged products, clearly derived from a carbene source, are often isolated in these reactions, sometimes to the exclusion of products of rearrangement. For example, thermolysis of the diazo-ketone (78) in aniline yields the anilide (81), the normal product of Wolff rearrangement, which is formed by a ring-contraction process. However, if the reaction is carried out in a good hydrogen-donor solvent such as benzyl alcohol, β-naphthol (82) is formed [24]. In this case, the intermediate carbene (79) abstracts hydrogen from the solvent before rearrangement (79 → 80) can occur.

(78) $\xrightarrow[-N_2]{\Delta}$ (79) $\xrightarrow{PhCH_2OH}$ (82)

(79) → (80) $\xrightarrow{PhNH_2}$ (81)

$PhCOC(Ph)N_2$ (83) $\xrightarrow[-N_2]{\Delta}$ $Ph_2C{=}C{=}O$ (84)

(85) $\xrightarrow[-N_2]{h\nu}$ (86)

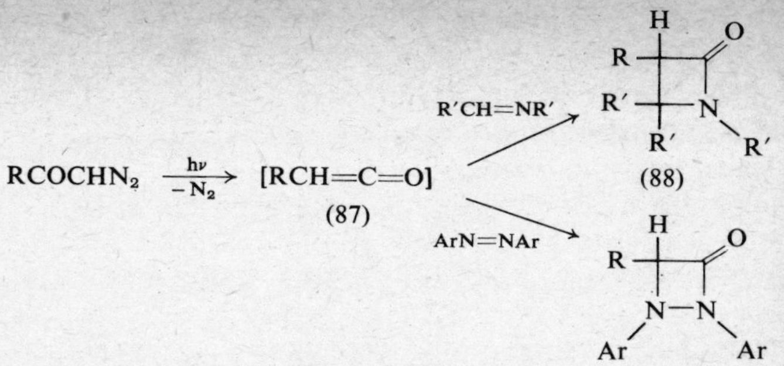

(87) (88) (89)

If the Wolff rearrangement is carried out in an inert solvent, the initial rearrangement product, a ketene (77), may be isolable in favourable cases, e.g. 83 → 84 [25] and 85 → 86 [26]. Aldoketenes (87) cannot be isolated since they dimerise spontaneously. However, they may be trapped by unsaturated compounds and so give synthetic routes to cyclic compounds [see W. Kirmse, *Carbene Chemistry*, Academic Press, New York (1964)]. For example, β-lactams (88) and diazetidinones (89) are formed when aldoketenes are generated photochemically in the presence of Schiff bases and arylazo-compounds respectively [27]. More commonly, the decomposition of diazo-ketones is conducted in solvents which react with ketones, and a range of compounds may be obtained in this way (see Scheme 4).

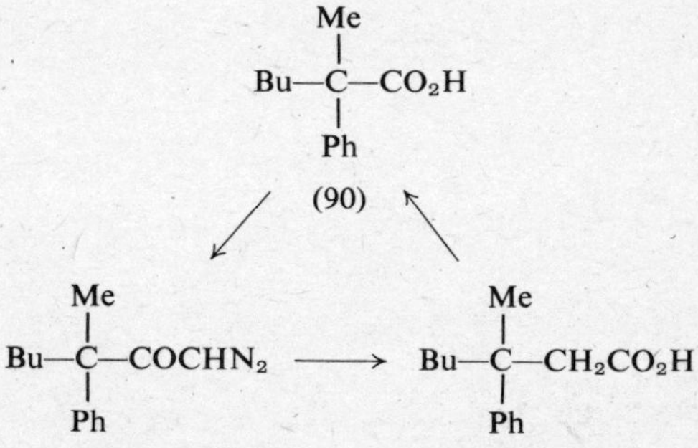

(90) (91) (92)

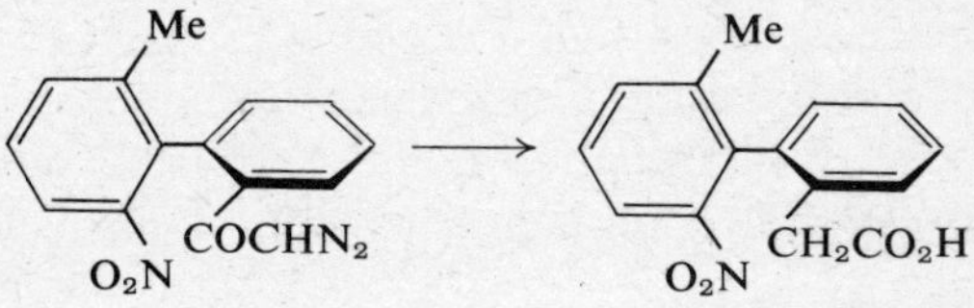

(93) (94)

The key step in the Wolff rearrangement (76 → 77) is thus the nucleophilic migration of a group to a carbenoid atom. In cases where the migrating carbon atom is asymmetrically substituted, the rearrangement proceeds with retention of configuration at this asymmetric centre. This was elegantly demonstrated [28] by Wallis who carried out the rearrangement of the diazo-ketone (91), prepared from optically active α-butylhydratropic acid (90). The rearrangement product (92) was reconverted to this acid by the Barbier–Wieland procedure (i.e., $RCH_2CO_2R \rightarrow RCH_2CR'_2OH \rightarrow RCH{=}CR'_2 \rightarrow RCO_2H$) and it was found that the original optical activity was almost completely (99·5 per cent) retained. In the 1,2-rearrangement step, therefore, the migrating carbon atom never becomes fully detached from the substrate.

The same conclusion was reached in a related experiment in which an optically active form of the biphenyl derivative (93) was rearranged. The product (94) was found to be optically active also. The optical activity of these compounds arises from the restricted rotation around the bond linking the two benzene rings, caused by steric interference by the *ortho*-substituents. If the migrating biphenylyl group had fully separated during the rearrangement this barrier to rotation would have been removed, and racemic product would have been isolated.

Scheme 5

$$RCOCHN_2 + B \rightleftharpoons BH^+ + RCO\bar{C}N_2 \quad (95)$$

$$RCO\bar{C}N_2 \xrightarrow{Ag^+} RCO\dot{C}N_2 + Ag \quad (96)$$

$$RCO\dot{C}N_2 \xrightarrow{-N_2} R\dot{C}{=}C{=}O \xrightarrow{RCOCHN_2} RCO\dot{C}N_2 + RCH{=}C{=}O$$

$$RCO\dot{C}N_2 \xrightarrow{-N_2} \textit{etc.}$$

Finally, the mechanism of the silver-catalysed Wolff rearrangement (*vide supra*) merits consideration. While the action of the catalyst has not been convincingly elucidated, evidence has been presented [29] in support of the free-radical chain process outlined in Scheme 5. Free-radical intermediates certainly appear to be involved in the process since the reaction is inhibited by hydroquinone, a radical quenching agent. Earlier workers had noted the reduction of silver ion to metallic silver during the course of such reactions, and Newman has proposed that Ag^+ functions as a one-electron oxidising agent which converts the conjugate base (95) of the diazo-ketone to a radical species (96). In conformity with the proposed mechanism, it was found that the presence of base (e.g. Et_3N) is necessary for reaction to occur. Further, since diazo-ketones that lack an acidic hydrogen at the migration terminus ($RCOCRN_2$) do not rearrange under the reaction conditions, generation of the conjugate base (95) appears to be necessary.

References

1. F. C. Whitmore, *J. Amer. Chem. Soc.*, 1932, **54**, 3274.
2. F. C. Whitmore and H. S. Rothrock, *J. Amer. Chem. Soc.*, 1932, **54**, 3431.
3. I. Dostrovsky and E. D. Hughes, *J. Chem. Soc.*, 1946, 166.
4. R. Fittig, *Annalen*, 1860, **114**, 54.
5. G. Wanger and W. Brickner, *Ber.*, 1899, **32**, 2302.
6. H. Meerwein and K. van Emster, *Ber.*, 1920, **53**, 1815; 1922, **55**, 2500.
7. N. Demjanov and M. Luschnikov, *J. Russ. Phys. Chem. Soc.*, 1903, **35**, 26.
8. G. Komppa and G. A. Nyman, *Ber.*, 1936, **69**, 334.
9. C. A. Bunton, T. Hadwick, D. R. Llewellyn, and Y. Pocker, *J. Chem. Soc.*, 1958, 403.
10. W. E. Bachmann *et al.*, *J. Amer. Chem. Soc.*, 1932, **54**, 1124; 1933, **55**, 3821; 1934, **56**, 170, 2081.
11. M. Stiles and R. P. Mayer, *J. Amer. Chem. Soc.*, 1959, **81**, 1497.
12. (a) M. Chérest, H. Felkin, J. Sicher, F. Šipoš, and M. Tichy, *J. Chem. Soc.*, 1965, 2513.
    (b) C. A. Bunton and M. D. Carr, *J. Chem. Soc.*, 1963, 5854.
13. P. von R. Schleyer, W. E. Watts, and C. A. Cupas, *J. Amer. Chem. Soc.*, 1964, **86**, 2722.
14. G. Brownlie, F. S. Spring, R. Stevenson, and W. S. Strachan, *J. Chem. Soc.*, 1956, 2419.
15. D. Y. Curtin and M. C. Crew, *J. Amer. Chem. Soc.*, 1955, **77**, 355.
16. H. I. Bernstein and F. C. Whitmore, *J. Amer. Chem. Soc.*, 1939, **61**, 1324.
17. D. J. Cram, *J. Amer. Chem. Soc.*, 1949, **71**, 3863.
18. S. Winstein *et al.*, *J. Amer. Chem. Soc.*, 1965, **87**, 3504, 3506.
19. J. C. Charlton, I. Dostrovsky, and E. D. Hughes, *Nature*, 1951, **167**, 986.
20. A. C. Cope, P. E. Burton, and M. L. Caspar, *J. Amer. Chem. Soc.*, 1962, **84**, 4855.
21. L. Wolff, *Annalen*, 1902, **325**, 134; 1912, **394**, 25.
22. F. Arndt and B. Eistert, *Ber.*, 1935, **68**, 200; 1936, **69**, 1074, 1805.
23. A. L. Wilds and A. L. Meader, *J. Org. Chem.*, 1948, **13**, 763.
24. P. A. S. Smith and W. L. Berry, *J. Org. Chem.*, 1961, **26**, 27.
25. L. I. Smith and H. H. Hoehn, *Org. Synth. Coll. Vol.* 3, 1955, 356.

26. O. Süs, *Annalen*, 1944, **556**, 65.
27. L. E. Horner *et al.*, *Chem. Ber.*, 1956, **89**, 2762, 2765.
28. J. F. Lane and E. S. Wallis, *J. Amer. Chem. Soc.*, 1941, **63**, 1674.
29. M. S. Newman and P. F. Beal, *J. Amer. Chem. Soc.*, 1950, **72**, 5162.

Review Articles

General reviews of carbonium-ion rearrangements

G. W. Wheland, *Advanced Organic Chemistry*, pp. 550–619; Wiley (New York), 1960.

J. March, *Advanced Organic Chemistry: Reactions, Mechanisms and Structure*, pp. 781–822; McGraw-Hill (New York), 1968.

Specialized reviews of carbonium-ion rearrangements:

Wagner–Meerwein and Pinacol rearrangements

Y. Pocker in *Molecular Rearrangements*, Part 1, pp. 1–26; ed. de Mayo, Interscience (New York), 1963.

Rearrangements in bridged bicyclic systems

J. A. Berson, loc. cit., pp. 111–232.

Rearrangements in small-ring compounds

R. Breslow, loc. cit., pp. 254–280.

Rearrangements of bicyclic terpenes

L. Ruzicka in *Perspectives in Organic Chemistry*, pp. 265–314, ed. Todd, Interscience (New York), 1956.

Rearrangements in solvolytic displacement reactions

A. Streitwieser, *Chem. Rev.*, 1956, **56**, 571.

Pinacol rearrangement

C. J. Collins, *Quart. Rev.*, 1960, **14**, 357.

Rearrangements involving non-classical carbonium ions

G. D. Sargent, *Quart. Rev.*, 1966, **20**, 301.

Wolff rearrangement

P. A. S. Smith in *Molecular Rearrangements*, Part 1, pp. 462–568; ed. de Mayo, Interscience (New York), 1963.

L. L. Rodina and I. K. Korobitsyna, *Russ. Chem. Rev.*, 1967, 260.

Arndt–Eistert procedure

W. E. Bachmann and W. S. Struve, *Org. Reactions*, 1942, **1**, 39.

3

Nucleophilic Rearrangements II: Migration from Carbon to Nitrogen

Molecular rearrangements involving the 1,2-shift of a carbon group from a carbon atom to an adjacent nitrogen atom encompass an extremely diverse collection of organic reactions. A large number of these transformations were first discovered in the late nineteenth and early twentieth centuries, a particularly fruitful period in the annals of organic chemistry, and engaged the attentions of some of the leading chemists of the day, notably Hofmann, Beckmann, Curtius, and Schmidt. Later research into the mechanisms of these reactions uncovered many common features, particularly with regard to the factors controlling the rearrangement sequence. In this connection, therefore, it is both appropriate and instructive to consider together groups of reactions embodying similar mechanistic details, and this approach will be adopted in the following discussion.

With few exceptions, all of the processes described in this chapter involve the movement of a carbon group with its pair of bonding electrons to an adjacent nitrogen atom that is either electron-deficient or potentially so. In this sense, these reactions may be categorised as nucleophilic or anionotropic rearrangements: they may be considered as the nitrogen analogues of the carbonium-ion rearrangements discussed in the previous chapter. The intrinsic instability associated with a first-row atom, such as nitrogen, that possesses an incomplete valence shell of six electrons provides the driving force for nucleophilic rearrangements of these types. However, it is not always easy to discern whether an electron-deficient nitrogen atom (—$\overset{+}{\underset{..}{N}}$— or —$\overset{..}{\underset{..}{N}}$) is actually created in the reaction or whether its formation is prevented by concomitant shift of a carbon group. It will be recalled that a similar problem of mechanism was posed in the interpretation of carbonium-ion rearrangements.

In the following discussion, the rearrangements are classified into three groups according to the nature of the rearranging species, which may be *formally* described as an *N*-substituted amine (R_2NX), imine ($R_2C{=}NX$), or amide ($RCONR'X$) or ions derived therefrom. The nitrogen substituent X in every case represents a leaving-group which is stable as an anion or a neutral molecule. Except in special cases where hydrogen shifts are involved, the discussion will be restricted to reactions in which a carbon group migrates.

Rearrangements of *N*-Substituted Amines

The rearrangement sequence in reactions of this type may be generalised as shown in Scheme 1. The migration of the group R may either follow or take place simultaneously with departure of the leaving-group X. As will become apparent, these two descriptions of the rearrangement represent the extremes of a spectrum of possible mechanisms that depend upon the degree of neighbouring-group participation by R in the elimination of X. The nitrogen atom at the migration terminus may be uncharged (1) or the rearrangement may involve the conjugate base (2), formed in basic solution by abstraction of a proton from the nitrogen atom. The rearrangement product, an

Scheme 1

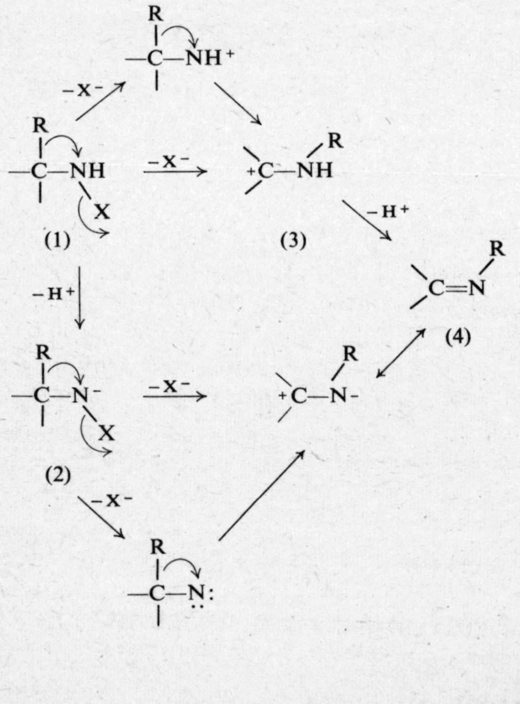

imine (4), may survive the reaction conditions and be capable of isolation. In many cases, however, the products obtained are derived from further reactions of the imine.

Alkyl and aryl azides. The electronic structure of organic azides may be represented as a resonance hybrid of the canonical forms 5A and 5B. These compounds, in particular the alkyl variety, are thermodynamically unstable molecules and potentially explosive in many instances. In acidic media the azido group becomes protonated, and the conjugate acids thus formed often undergo spontaneous rearrangement. It should be emphasised at this point, however, that the extent of the protonation will be low, even in highly concentrated acids (cf. pK_a of $H_2N_3^+$ = $-6{\cdot}2$). In principle, addition of a proton to either end atom of the linear N_3 fragment can be envisaged, to yield different cations (6 and 7). However, the rearrangements observed for azides in acidic media become intelligible only if the reacting species has the structure 6 (i.e. 1; X = N_2^+), in which the substituted nitrogen atom has accepted a proton. Such a process causes a weakening of the bond between this nitrogen atom and its neighbour, and elimination of a nitrogen molecule by heterolytic fission of this bond is facilitated. For this reason, the rearrangement of azides in the absence of acid usually requires much higher temperatures than the corresponding acid-catalysed reaction.

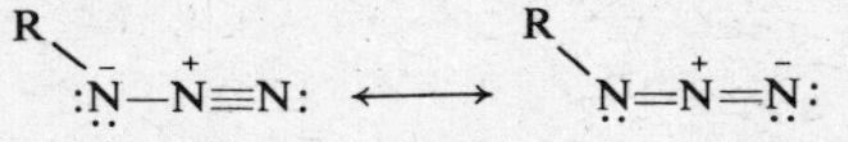

$$R{-}\ddot{\overset{-}{N}}{-}\overset{+}{N}{\equiv}N: \longleftrightarrow R{-}\ddot{N}{=}\overset{+}{N}{=}\ddot{\overset{-}{N}}:$$

(5A) (5B)

$$R{-}\ddot{N}H{-}\overset{+}{N}{\equiv}N: \qquad R{-}\ddot{N}{=}\overset{+}{N}{=}\ddot{N}H$$

(6) (7)

The first report [1] of the rearrangement of an organic azide in acidic media can be traced to Curtius, who described the conversion of benzhydryl azide (8; Ar = Ph) to benzylidene aniline (9; Ar = Ph) on treatment with concentrated sulphuric or hydrochloric acids. This reaction was found to be general for diarylalkyl azides (8) and is also adaptable to the synthesis of phenanthridines (11) and other fused aromatic heterocycles (e.g. 13) by ring expansion of the corresponding azides (10 and 12 respectively) [2]. In the last example, it will be noted that the product of phenyl migration (14) is also formed. We shall return to the question of competitive migration later.

$$\underset{(8)}{Ar_2CHN_3} \xrightarrow{H^+} Ar_2CHNHN_2^+ \xrightarrow[-H^+]{-N_2} \underset{(9)}{ArCH{=}NAr}$$

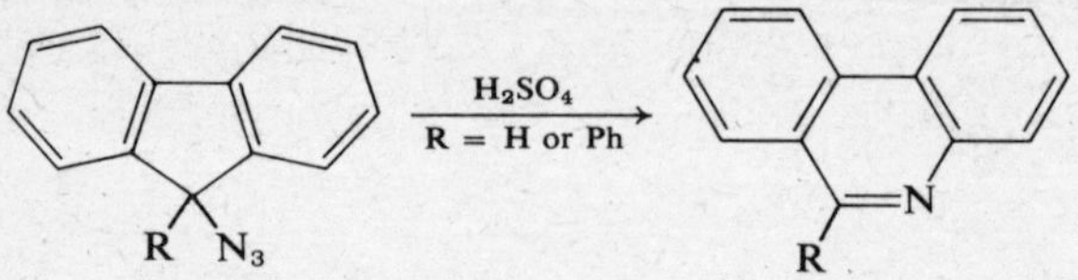

(10) $\xrightarrow[R\,=\,H\text{ or }Ph]{H_2SO_4}$ (11)

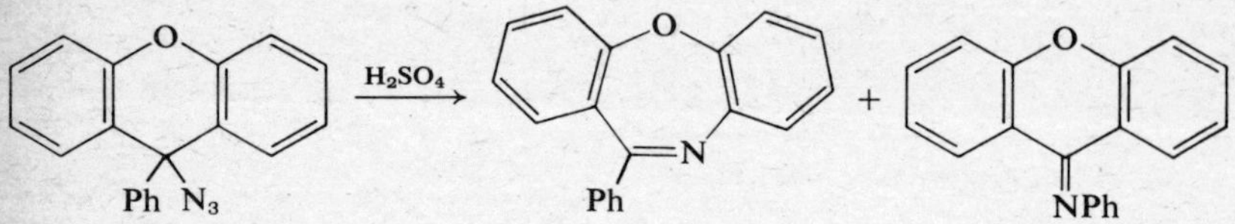

(12) $\xrightarrow{H_2SO_4}$ (13) + (14)

Simple triarylmethyl azides (Ar_3CN_3), on the other hand, are usually inert to acid-catalysed rearrangement and yield the corresponding alcohols (Ar_3COH) when treated with protic acids. This behaviour may be contrasted with the thermal reactions of these compounds which are rearranged to anils ($Ar_2CH{=}NAr$) on heating to about 200° in the absence of acid [3].

Treatment of alkyl azides with protic acids also leads to rearrangement. The imino-compounds formed, however, are unstable in the acidic media employed for the reaction, and are hydrolysed to the corresponding carbonyl compound and amine (or ammonia). For example, ethyl azide (15) reacts with strong sulphuric acid to give a number of products whose formation can be interpreted in terms of both methyl and hydrogen migration, and hydrolysis of the resulting imines (16 and 17 respectively) [4].

$$\underset{(15)}{CH_3CH_2N_3} \xrightarrow{H^+} CH_3CH_2NHN_2^+ \nearrow \underset{(16)}{[CH_2{=}NCH_3]} \longrightarrow CH_2O + CH_3NH_2$$

$$\searrow \underset{(17)}{[CH_3CH{=}NH]} \longrightarrow CH_3CHO + NH_3$$

Ring-expansion reactions have also been observed for cycloalkyl azides. This reaction has been put to ingenious use, for example, in the preparation of the pyridinophanes (22 and 23) from the bicycloalkene (18). Rearrangement of the intermediate azide conjugate acid (19) can proceed by either of two independent ring-expansion pathways to give the bicyclic imines (20 and 21) which were converted to the products by dehydrogenation [5].

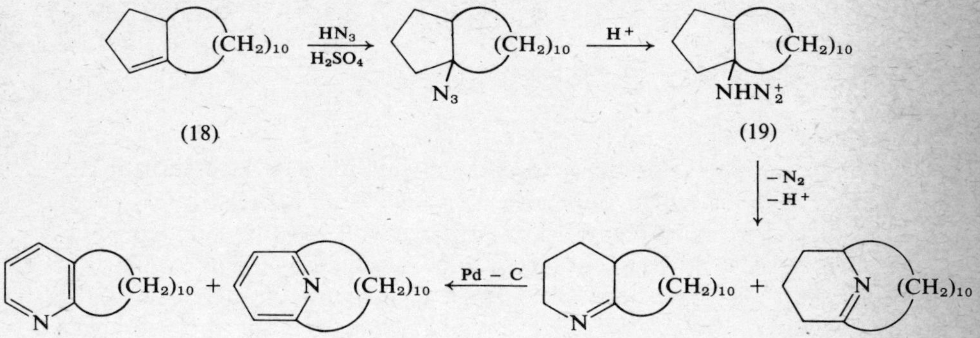

$(CH_2)_{10}$ $\xrightarrow[H_2SO_4]{HN_3}$ $(CH_2)_{10}$ N_3 $\xrightarrow{H^+}$ $(CH_2)_{10}$ NHN_2^+

(18) (19)

$\downarrow -N_2, -H^+$

$(CH_2)_{10}$ + N $(CH_2)_{10}$ $\xleftarrow{Pd-C}$ $(CH_2)_{10}$ + N $(CH_2)_{10}$

(22) (23) (20) (21)

The reactions of alkenes or alcohols with hydrogen azide in the presence of strong acids (usually H_2SO_4) to give imino-compounds, exemplified in the foregoing reaction sequence, are commonly referred to as Schmidt reactions, although this designation is more usually applied to the corresponding reactions of carboxylic acids, aldehydes, and ketones (see next section). The intermediacy of azides in these transformations is well established and, in fact, azides are often isolated together with rearrangement products from Schmidt reactions. The function of the acid catalyst in the process is thus two-fold, i.e., to convert the alcohol or alkene to a carbonium ion that leads to azide, and to promote rearrangement of the azide. The sequence of events may thus be described as shown in Scheme 2. Examples of the application of the Schmidt reaction to the synthesis of cyclic and acyclic imines are provided by the transformations 24 → 25 [6] and 26 → 27 [7].

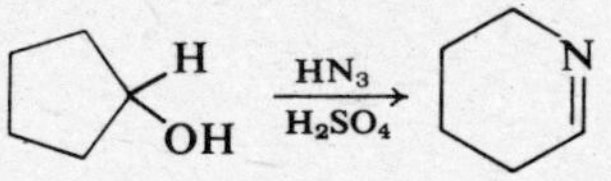

H OH $\xrightarrow[H_2SO_4]{HN_3}$ N

(24) (25)

$$Ph_2C{=}CH_2 \xrightarrow[H_2SO_4]{HN_3} PhC(Me){=}NPh \longrightarrow PhCOMe + PhNH_2$$

(26) (27)

Scheme 2

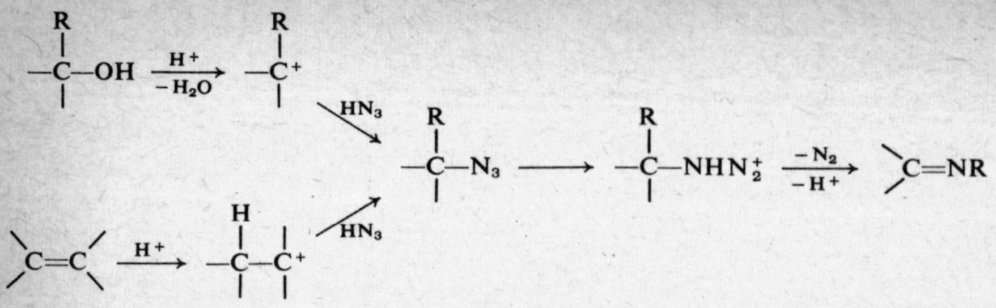

$$\mathrm{R{-}\overset{|}{\underset{|}{C}}{-}OH \xrightarrow[-H_2O]{H^+} R{-}\overset{|}{\underset{|}{C}}{}^{+} \xrightarrow{HN_3} R{-}\overset{|}{\underset{|}{C}}{-}N_3 \longrightarrow R{-}\overset{|}{\underset{|}{C}}{-}NHN_2^+ \xrightarrow[-H^+]{-N_2} {>}C{=}NR}$$

$$\mathrm{{>}C{=}C{<} \xrightarrow{H^+} H{-}\overset{|}{\underset{|}{C}}{-}\overset{|}{\underset{|}{C}}{}^{+} \xrightarrow{HN_3} {-}\overset{|}{\underset{|}{C}}{-}N_3}$$

The mechanism of the azide rearrangement has been thoroughly studied. The rate of evolution of nitrogen from azides on treatment with acids has been found to be proportional both to the azide concentration and to the acid strength of the medium. The conclusion appears inescapable that the conjugate acid (6) of the azide is the rearranging species. Evidence has also been obtained for the acid-catalysed reaction that shift of the migrating-group and elimination of a nitrogen molecule occur in a concerted fashion. For example, in the case of aryl migration (8 → 9), the rate of the process is accelerated by substituents attached to the aryl ring which can donate electrons (e.g. *p*-OMe) and retarded by electron-withdrawing groups (e.g. *p*-NO_2). In other words, the height of the free-energy barrier to the rearrangement is lowered when the migrating group is able effectively to disperse the positive charge and thereby stabilise the transition state (28) for the process.

In contrast, the thermal rearrangement of azides is relatively insensitive to substituent effects in the case of aryl migration [8]. It

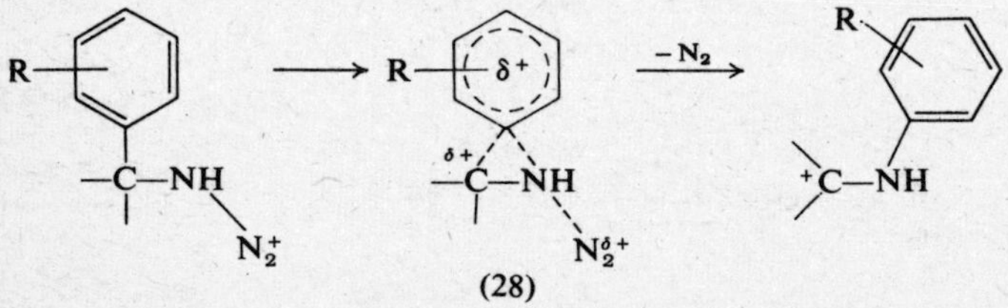

$$\mathrm{R{-}C_6H_4{-}\overset{|}{\underset{|}{C}}{-}NH{-}N_2^+ \longrightarrow [R{-}C_6H_4^{\delta+}\cdots\overset{\delta+}{C}{-}NH\cdots N_2^{\delta+}] \xrightarrow{-N_2} {}^{+}C(-NH{-}C_6H_4{-}R)}$$

(28)

$$\mathrm{R{-}C_6H_4{-}\overset{|}{\underset{|}{C}}{-}N^-{-}N_2^+ \xrightarrow{-N_2} R{-}C_6H_4{-}\overset{|}{\underset{|}{C}}{-}\ddot{N}{:} \longrightarrow {>}C{=}N{-}C_6H_4{-}R}$$

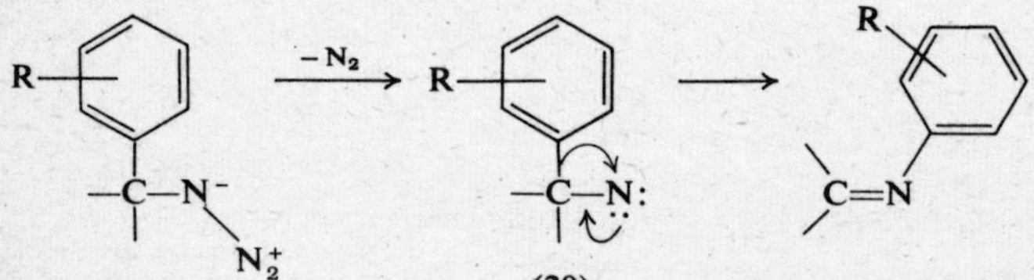

(29)

seems likely that a different mechanism operates, and a two-step process that involves the generation of an intermediate nitrene (29) appears probable.

As a consequence of the reaction mechanism, all groups do not migrate with equal ease in the acid-catalysed azide rearrangement (cf. behaviour of carbonium ions; Chapter 2). For example, the anil (31) is the sole product of the reaction of the azide (30) with sulphuric acid, the anisyl group migrating to the exclusion of phenyl or methyl shifts [9]. A similar effect is apparent in the rearrangement of substituted fluorenyl azides (32) which undergo ring expansion to give a mixture of isomeric phenanthridines [10]. In these reactions, preferential migration of the group most capable of sustaining positive charge is observed.

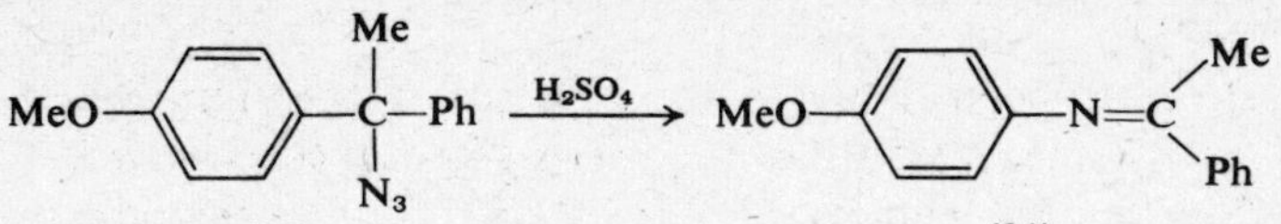

MeO—C₆H₄—C(Me)(N₃)—Ph $\xrightarrow{H_2SO_4}$ MeO—C₆H₄—N=C(Me)Ph

(30) (31)

$\xrightarrow{H_2SO_4}$ +

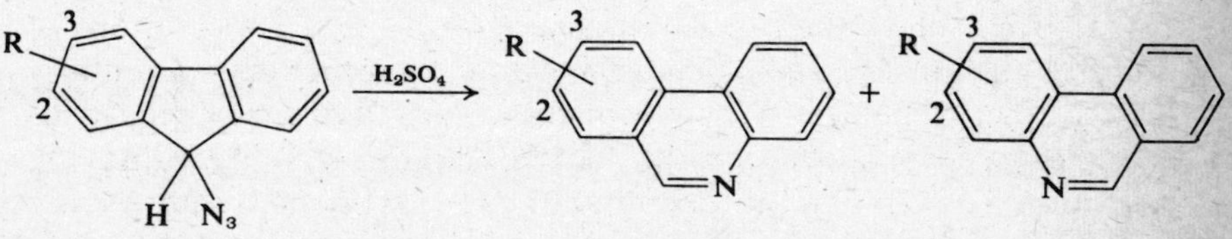

(32) (33) (34)

R = 2–Me ⟶ 47% (33) + 53% (34)
R = 3–Me ⟶ 0% (33) + 100% (34)
R = 2–NO_2 ⟶ 97% (33) + 3% (34)
R = 3–NO_2 ⟶ 94% (33) + 6% (34)

Conformational factors are often important in the determination of migratory behaviour. Consider the geometry of the conjugate acid (35) in which three groups of disparate size (S = small; M = medium; L = large) may potentially migrate to the adjacent nitrogen atom. Various rotameric conformations are obviously possible for a cation of this type (e.g. 35A, 35B, and 35C). The most stable (i.e. lowest energy and highest population) arrangement is that in which steric repulsions are minimised and in which the N_2^+ group is flanked by the smallest groups (S and M) attached to the neighbouring carbon atom (i.e., conformation 35A).

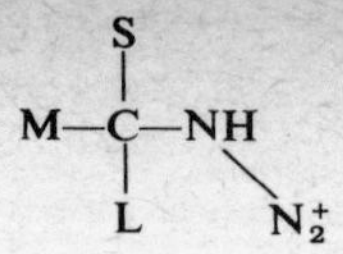

(35)

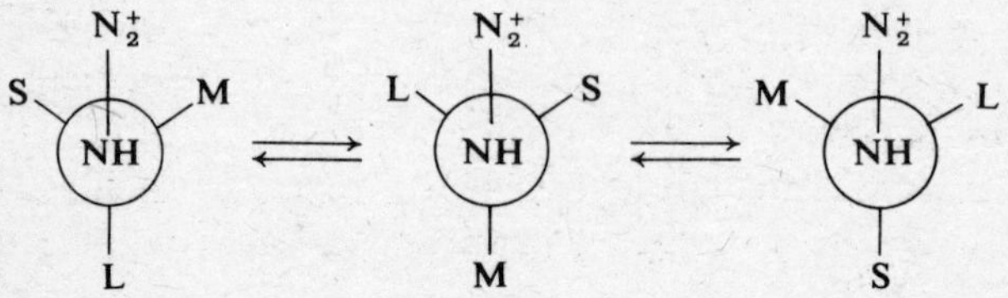

(35A) ⇄ (35B) ⇄ (35C)

Reference has already been made to the evidence that migration occurs synchronously with departure of the leaving-group (N_2) in this 1,2-shift. A concerted rearrangement of this nature requires that the migrating- and leaving-groups be arranged in an *anti*-periplanar relationship, and it is obvious that a different group would perforce migrate for each of the three conformers shown (e.g. L in conformation 35A). It must be remembered, however, that the relative proportions of the isomeric products formed are not necessarily related to the relative populations of the various conformers preceding the migration step *unless* rotation around the C—NH bond is appreciably slower than the rate of rearrangement. If rotational equilibration (35A ⇄ 35B ⇄ 35C) occurs faster than the rate of rearrangement, the distribution of isomeric products will be controlled only by the free-energy differences between the conformer substrates and the corresponding transition states (Curtin–Hammett principle). As mentioned previously, the height of these energy barriers is sensitive to the nature of the migrating-group and its ability to delocalise positive charge.

The actual course followed in a particular acid-catalysed azide rearrangement therefore reflects an interplay and a balancing of these effects. In examples where there is competition between alkyl and aryl migration, for example, shift of an aryl group is favoured by electronic considerations (e.g. 26 → 27; cf. 28) but this preference may be outweighed by adverse steric factors [11]. Thus, Schmidt reaction of the alcohol (36; R = Me) gives predominantly the anil (37; R = Me), phenyl migration occurring in preference to methyl migration (i.e., via 39). The major product of the rearrangement of the isopropyl analogue (36; R = $CHMe_2$), on the other hand, is the anil (38; R = $CHMe_2$) that results from shift of the relatively bulky isopropyl group (i.e., via 40).

$$\text{Ph—C(R)(Me)—OH} \xrightarrow[H_2SO_4]{HN_3} \text{PhN=C(R)(Me)} + \text{RN=C(Ph)(Me)}$$

(36) (37) (38)

(39) Newman projection: N_2^+ (top), Me, Me, NH (centre), Ph (bottom)

(40) Newman projection: N_2^+ (top), Ph, Me, NH (centre), $CHMe_2$ (bottom)

Hydroxylamines and related compounds. Other reactions are known which bear close similarity to the azide rearrangement and may be described by the same general mechanism (1 → 3 → 4). For example, triarylmethyl hydroxylamines (41) are converted to anils (43) on treatment with phosphorus pentachloride in ether [12]. The key step in this transformation may be formulated as elimination of a stable phosphorus oxychloride molecule and chloride ion from the intermediate (42), with concomitant aryl migration.

A number of related base-catalysed rearrangements have also been found. Compounds of the type 44 (R = H; X = Cl, OCOPh, etc.) rearrange on treatment with alkoxide or hydroxide to give the corresponding anils (43). Since the methylated analogues (44; R = Me) do not rearrange under similar conditions, the rearranging species in the former reaction is established as the conjugate base (45), formed by abstraction of a proton from the NH group of the precursor.

$$\underset{(41)}{Ar_3CNHOH} \xrightarrow[-HCl]{PCl_5} \underset{(42)}{Ar_2C(Ar)\text{—}NH(OPCl_4)} \xrightarrow[-Cl^-]{-POCl_3} Ar_2\overset{+}{C}\text{—}NHAr \xrightarrow{-H^+} \underset{(43)}{Ar_2C{=}NAr}$$

$$\underset{(44)}{Ar_3CN(R)(X)} \xrightarrow[\substack{-H^+ \\ (R\,=\,H)}]{RO^-} \underset{(45)}{Ar_2C(Ar)\text{—}\bar{N}\text{—}X} \xrightarrow{-X^-} \underset{(43)}{Ar_2C{=}NAr}$$

Finally, the thermal rearrangement of aryl azides (ArN_3) merits comment. For example, the azepine derivative (48) is formed when a solution of phenyl azide in aniline is heated [13]. The rate of this reaction is first-order in azide concentration and relatively insensitive to the presence of substituents in the benzene ring. The formation of an intermediate nitrene appears likely. This species (46) may undergo an intramolecular insertion reaction to give the highly strained bicyclic imine (47), which is quickly destroyed by the solvent (aniline) to give the seven-membered ring product.

$$PhN^{-}\!-\!N_2^{+} \xrightarrow{-N_2} Ph\ddot{N}: \longrightarrow (47) \xrightarrow{PhNH_2} [\ldots \overset{+}{N}H_2Ph] \longrightarrow (48)\ NHPh$$

(46) (47) (48)

Rearrangements of *N*-Substituted Imines

Several organic reactions are known in which the key step involves rearrangement of an imino species (49) whose nitrogen atom bears a good leaving-group X that is stable as an anion or a neutral molecule. The product of the rearrangement (50) may either recombine with the eliminated group X, if this possesses nucleophilic properties, or alternatively react with other molecules (M) present in the reaction medium. The products (51 and 52) formed in these ways may then undergo further change under the reaction conditions.

Perhaps the best known example of a reaction of this type is the **Beckmann rearrangement** of oximes to amides, first reported in 1886 [14]. As usually carried out, an ethereal solution of the oxime (49; X = OH) is treated with phosphorus pentachloride and, when the vigorous reaction has subsided, the product is precipitated by pouring the reaction mixture over ice. The rearrangement may be effected by a range of catalysts, including strong acids (e.g. conc. H_2SO_4; polyphosphoric acid) and acid chlorides (e.g. $SOCl_2$; RCOCl; RSO_2Cl). The reaction may be applied to the oximes of both aryl and alkyl ketones and to cyclic ketoximes. Aldoximes, on the other hand, normally suffer dehydration to nitriles (RCH=NOH → RCN) under the conditions of the rearrangement (*vide infra*) and examples of the formation of formamides (HCONHR) from aldoximes are rare. The application and use of the Beckmann rearrangement in organic synthesis are illustrated by the transformations: 53 → 54, 55 [15], 56 → 57 [16], and 58 → 59 [17].

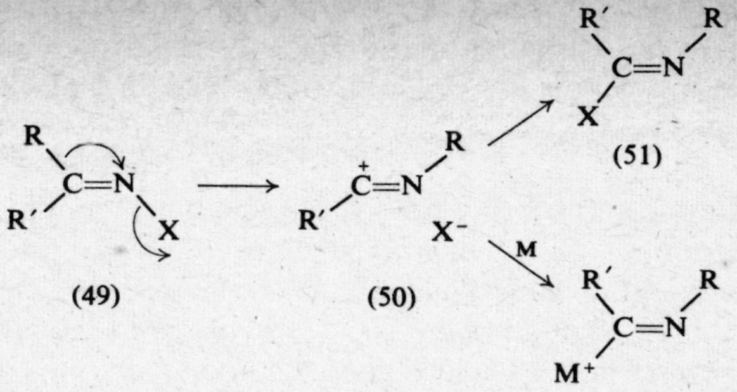

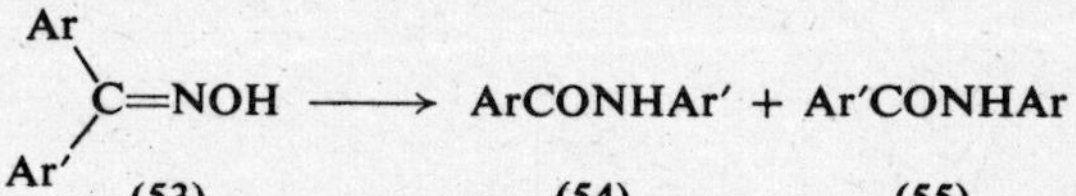

Ar(Ar′)C=NOH ⟶ ArCONHAr′ + Ar′CONHAr

(53) (54) (55)

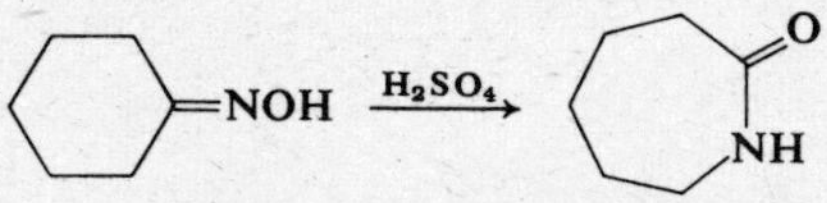

=NOH $\xrightarrow{H_2SO_4}$ NH, O

(56) (57)

NOH, NOH $\xrightarrow{(H_3PO_4)_n}$ O, H, N, N, O, H

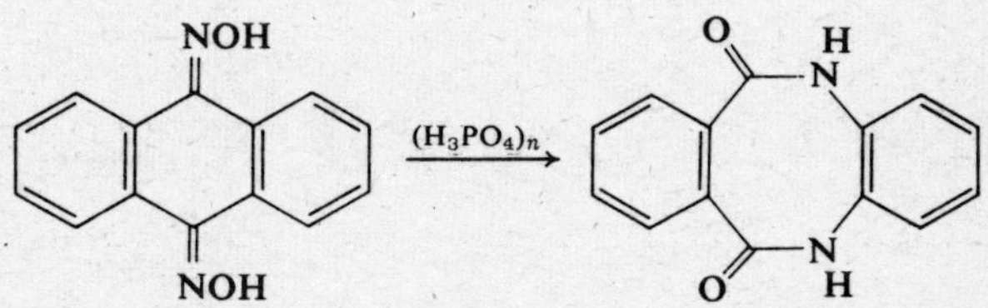

(58) (59)

The history of the early attempts to unravel the mechanism of the Beckmann rearrangement constitutes an interesting chapter in the development of organic chemistry. Various proposals were advanced, later only to be proved untenable as fresh information came to light. The problem faced in these original studies was complicated by the geometrical isomerism associated with unsymmetrical ketoximes that can exist in different stereochemical forms (60 and 61). Without the benefits of modern physical and spectroscopic techniques, unambiguous assignment of the correct structure to a particular oxime proved no straightforward business. The problem was further confused by the ready equilibration (60 ⇄ 61) of geometrically isomeric oximes in the presence of acid catalysts. Since the conditions of the rearrangement require catalysts of just this type, the amide produced in a particular reaction may actually be derived from the geometrical isomer of the oxime originally used. For example, rearrangement of either form (62 or 63) of *o*-methoxybenzophenoxime gives a mixture of amides (64 and 65) in which the product of migration of the anisyl group (i.e. 64) greatly predominates [18].

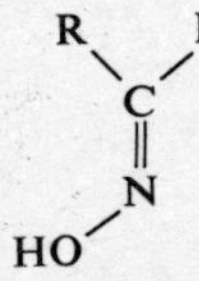

R R′ C=N HO
(60)

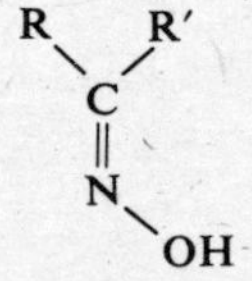

R R′ C=N OH
(61)

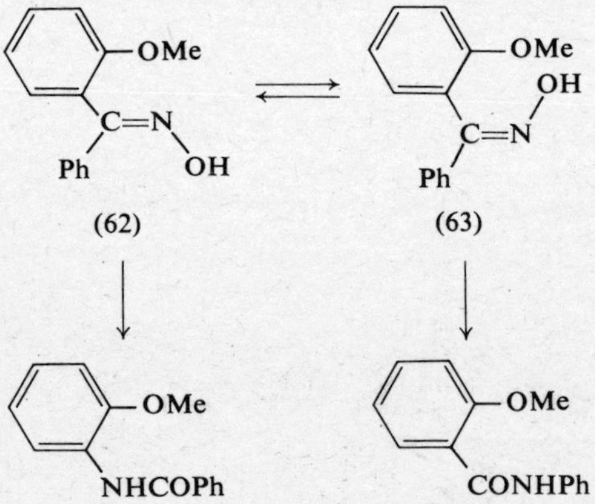

OMe C=N Ph OH
(62)

⇄

OMe OH C=N Ph
(63)

↓

↓

OMe NHCOPh
(64)

OMe CONHPh
(65)

Oximes themselves do not rearrange spontaneously, and the function of the catalyst is to convert the hydroxyl group to a better leaving-group. For example, the Beckmann rearrangement is catalysed by arenesulphonyl chlorides which react with the oxime giving the corresponding sulphonate (66). Esters of this type have in fact been isolated and shown to rearrange spontaneously in solution. The product of the reaction in an inert solvent is the imidyl derivative (67) which may be converted to amide by hydrolysis. If the reaction is conducted in a nucleophilic solvent, the normal course may be diverted. For example, imino-ethers (68) are formed when the sulphonates are decomposed in ethanol. It has also been established [19] that the rate of the rearrangement of oxime esters (RCH=NOA; A = COR′, SO_2R', etc.) is proportional to the acid strength of the esterifying acids. Consequently, sulphonyl chlorides are much more effective catalysts than carboxylic acid chlorides for the rearrangement of oximes, since the sulphonate leaving-group ($R'SO_3^-$) is intrinsically much more stable than the carboxylate anion ($R'CO_2^-$) and is thus more readily eliminated from the ester intermediate (cf. RSO_3H is a much stronger acid than RCO_2H). Similarly, it has been found that the *p*-bromobenzenesulphonates (66; Ar = *p*-BrC_6H_4) rearrange at much faster rates than the corresponding *p*-toluenesulphonates (66; Ar = *p*-MeC_6H_4) since the brosylate anion is a better leaving-group than the tosylate anion.

$$R_2C{=}NOH \xrightarrow[-HCl]{ArSO_2Cl} \underset{(66)}{R_2C{=}NOSO_2Ar}$$

$$\underset{(68)}{R(OEt)C{=}NR} \xleftarrow[-H^+]{EtOH} \left[R\overset{+}{C}{=}NR \quad \overset{-}{O}SO_2Ar\right] \longrightarrow \underset{(67)}{R(OSO_2Ar)C{=}NR} \xrightarrow{H_2O} RCONHR$$

It is now well established that the Beckmann rearrangement proceeds by exclusive migration of the group located *trans* to the leaving-group in the substrate (i.e. R and not R′ in 49). Thus, the amide (69) is the sole product of rearrangement of the oxime (70) whose structure was established [20] by base-induced cyclisation to the isoxazole (71).

There is also good evidence to support the contention that migration occurs synchronously with departure of the leaving-group in most cases, since the rate of the process is extremely sensitive to the electronic capabilities of the migrating-group. As in the case of the

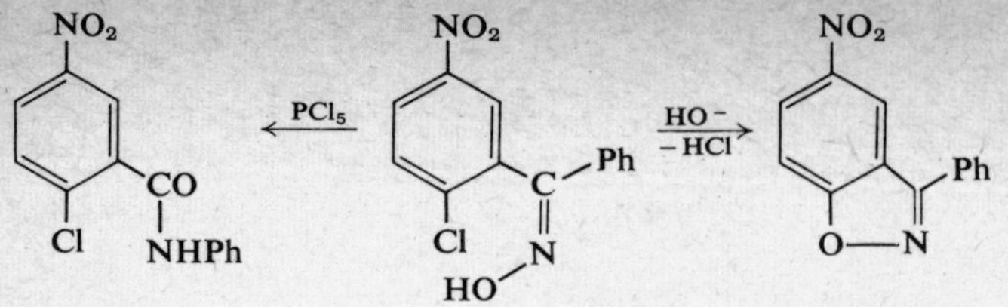

(69) (70) (71)

azide rearrangement (see preceding section), electron-withdrawing substituents attached to a migrating aryl group retard the rate of the reaction. The presence of electron-donating substituents, on the other hand, has an accelerating influence, since the migrating-group is then better able to render anchimeric assistance to the elimination of the leaving-group via the transition state (72).

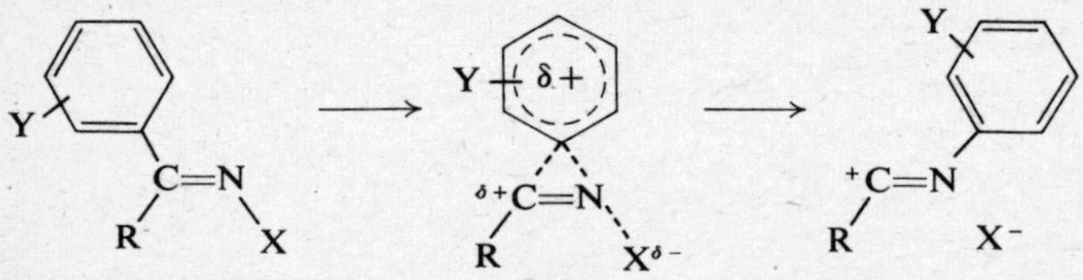

(72)

Besides these neighbouring-group phenomena, there are several other consequences of the operation of a mechanism of the type suggested.

(a) Since the rate-determining step in the reaction involves separation of charged species (i.e. 49 → 50), the rearrangement should proceed faster in polar solvents which facilitate ionisation. This has been found to be the case. The reaction is almost completely suppressed in hydrocarbon solvents (non-polar in character) but takes place readily in solvents such as ether, alcohol, and acetonitrile.

(b) For substrates in which the migrating carbon atom is asymmetrically substituted, rearrangement should occur with retention of configuration at this centre. In support of this conclusion, it has been found [21] that rearrangement of an optically active form of the oxime (73) to the amide (74) occurs with about 99 per cent retention of configuration at the asymmetric carbon atom.

(c) Since the oxygen atom of the oxime is eliminated during the rearrangement sequence, the carbonyl oxygen atom of the amide product may derive from solvent molecules. This has been demonstrated by isotopic labelling experiments. Thus, the benzanilide product

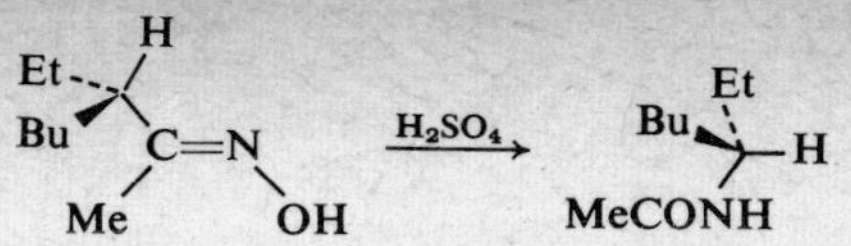

$$\text{Et, Bu, H}\!-\!\text{C}(\text{Me})\!=\!\text{N}\!-\!\text{OH} \xrightarrow{H_2SO_4} \text{Bu(Et)(H)C}\!-\!\text{NHCOMe}$$

(73) (74)

(54; Ar = Ar′ = Ph) produced in the Beckmann rearrangement of benzophenoxime (53; Ar = Ar′ = Ph) in the presence of $H_2{}^{18}O$ was found to contain labelled oxygen, although the amide itself does not exchange its oxygen atom in the presence of $H_2{}^{18}O$ under the conditions of the reaction.

(d) Since the migrating-group never becomes fully detached from the substrate during the rearrangement (cf. 72), the migration should proceed in a strictly intramolecular sense. Although there is evidence to suggest that this conclusion is correct for many Beckmann transformations [e.g., see (b) above], intermolecular migrations have also been observed. For example, treatment of a mixture of the oximes (75 and 76) with polyphosphoric acid affords two products (79 and 80) that result from 'crossover' migration, in addition to the expected products (77 and 78) of intramolecular rearrangement [22].

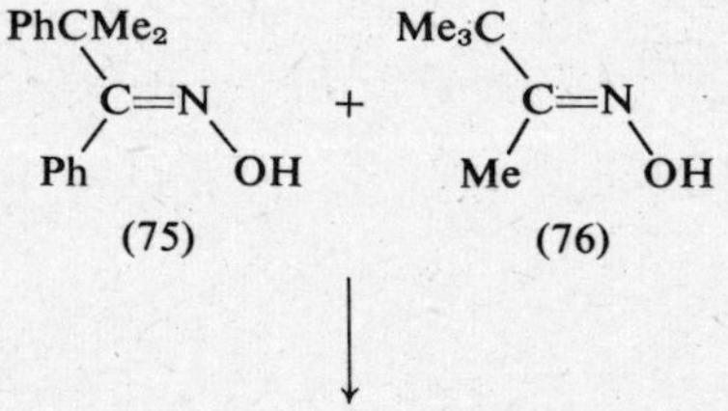

$$\text{PhCMe}_2(\text{Ph})\text{C}\!=\!\text{N}\!-\!\text{OH} \ (75) \ + \ \text{Me}_3\text{C}(\text{Me})\text{C}\!=\!\text{N}\!-\!\text{OH} \ (76)$$

$$\downarrow$$

$$\underset{(77)}{\text{PhCONHCMe}_2\text{Ph}} + \underset{(78)}{\text{MeCONHCMe}_3} + \underset{(79)}{\text{PhCONHCMe}_3} + \underset{(80)}{\text{MeCONHCMe}_2\text{Ph}}$$

Since the migrating groups in this reaction (i.e. cumyl and t-butyl) represent species relatively stable as carbonium ions, fragmentation of the substrates followed by indiscriminate recombination of the liberated cations and nitriles may occur as indicated in Scheme 3. The formation of benzonitrile (82) from pivalophenoxime (81), in fact, has been observed [23] and the dehydration of oximes (*vide supra*) represents a special case of reactions of this type in which a proton is eliminated.

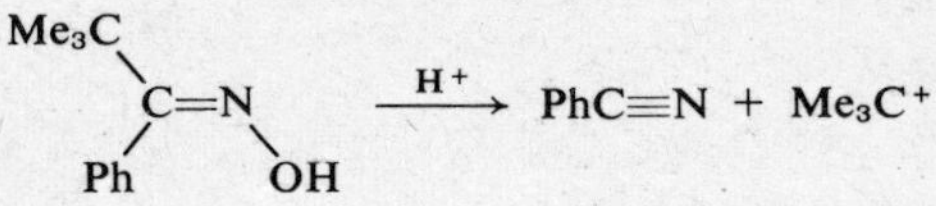

$$\underset{(81)}{\text{Me}_3\text{C}(\text{Ph})\text{C}\!=\!\text{N}\!-\!\text{OH}} \xrightarrow{H^+} \underset{(82)}{\text{PhC}\!\equiv\!\text{N}} + \text{Me}_3\text{C}^+$$

Scheme 3

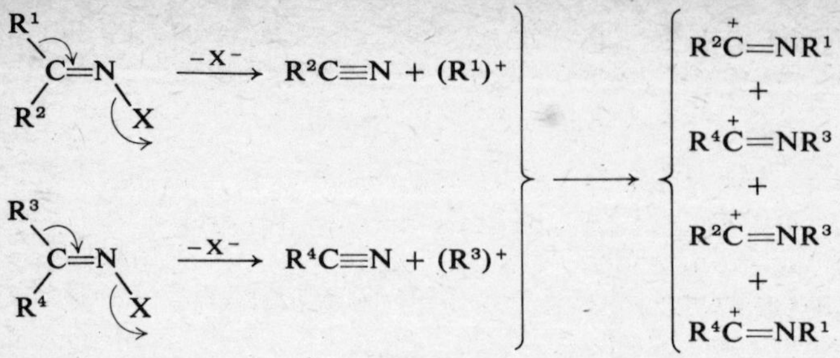

$$R^1R^2C{=}N{-}X \xrightarrow{-X^-} R^2C{\equiv}N + (R^1)^+$$

$$R^3R^4C{=}N{-}X \xrightarrow{-X^-} R^4C{\equiv}N + (R^3)^+$$

$$\longrightarrow \begin{cases} R^2\overset{+}{C}{=}NR^1 \\ + \\ R^4\overset{+}{C}{=}NR^3 \\ + \\ R^2\overset{+}{C}{=}NR^3 \\ + \\ R^4\overset{+}{C}{=}NR^1 \end{cases}$$

Fragmentation reactions of the type indicated in Scheme 3 are commonly encountered, therefore, in cases where a stable cation may be eliminated. For example, the oxime (83) of benzoin reacts with benzenesulphonyl chloride to give benzonitrile (82) and benzaldehyde. Under similar conditions, its geometrical isomer (84) undergoes a normal Beckmann 1,2-shift of a phenyl group, but the resulting cation (85) fragments to phenyl isocyanide (86) and benzaldehyde [24]. Reactions of this type are often extremely useful for the synthesis of compounds which are difficult to obtain by other routes, for example the preparation of the diphenic derivative (88) from the mono-oxime (87) of phenanthraquinone.

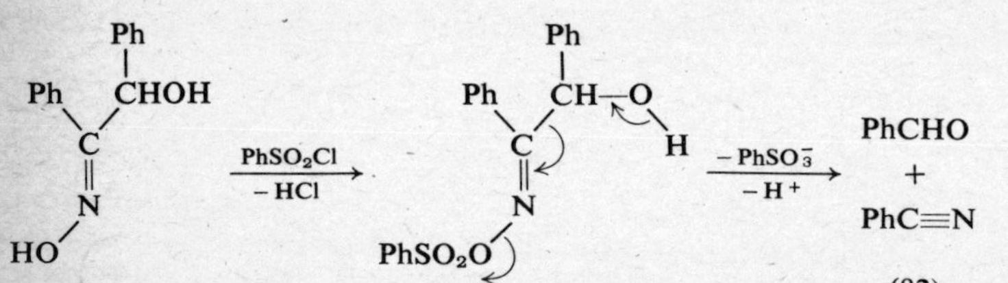

$$\underset{(83)}{Ph(PhCHOH)C{=}N{-}OH} \xrightarrow[-HCl]{PhSO_2Cl} Ph(PhCH{-}O{-}H)C{=}N{-}OSO_2Ph \xrightarrow[-H^+]{-PhSO_3^-} PhCHO + \underset{(82)}{PhC{\equiv}N}$$

$$\underset{(84)}{Ph(PhCHOH)C{=}N{-}OH} \xrightarrow[-HCl]{PhSO_2Cl} Ph(PhCHOH)C{=}N{-}OSO_2Ph \xrightarrow{-PhSO_3^-} \underset{(85)}{(PhCH{-}O{-}H)\overset{+}{C}{=}N{-}Ph} \xrightarrow{-H^+} PhCHO + \underset{(86)}{Ph\ddot{N}{=}C{:} \leftrightarrow Ph\overset{+}{N}{\equiv}\overset{-}{C}{:}}$$

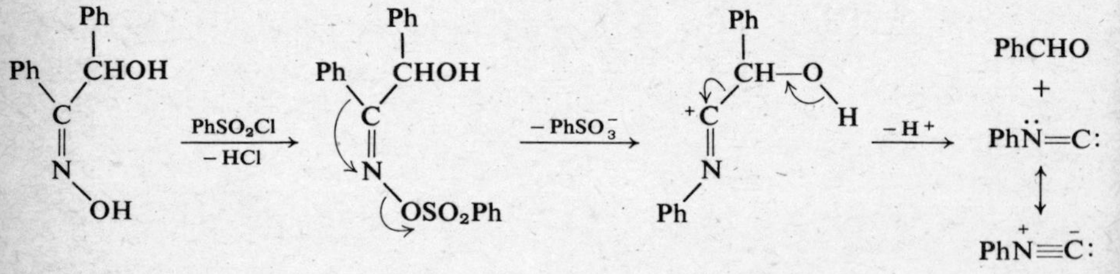

$$\text{(87)} \xrightarrow{H_2SO_4} \text{(88)}$$

(87) O, NOH (88) HO_2C CN

Schmidt reactions. As a result of his belief that a free imide radical (NH) was formed when hydrogen azide was decomposed in sulphuric acid, Schmidt attempted to trap this intermediate by carrying out the reaction in the presence of benzophenone (Ph_2CO) and observed [25] the production of benzanilide (PhCONHPh). The significance of this reaction for organic synthesis was quickly recognised when it was found that it could be applied to aldehydes, ketones, and carboxylic acids generally, giving high yields of products according to the equations:

$$RCHO + HN_3 \xrightarrow{-N_2} RC{\equiv}N + RNHCHO$$

$$RCOR' + HN_3 \xrightarrow{-N_2} RCONHR' + R'CONHR$$

$$RCO_2H + HN_3 \xrightarrow{-N_2} RNH_2 + CO_2$$

The Schmidt reaction is particularly useful for the preparation of both aliphatic and aromatic amines from the corresponding carboxylic acids, and offers the advantage over the related Hofmann and Curtius procedures (see next section) that the acid can be used directly without the necessity of preparing a suitable derivative. Examples illustrative of the application of the Schmidt reaction are provided by the transformations 89 → 90 [26], 91 → 92 [27], and 93 → 94 [28].

Schmidt's postulate of a reactive NH intermediate has now been discredited. The currently accepted version of the mechanism of the

$$3,5\text{-}(O_2N)_2C_6H_3CO_2H \xrightarrow[H_2SO_4]{NaN_3} 3,5\text{-}(O_2N)_2C_6H_3NH_2$$

(89) (90)

$$PhCH{=}CHCOMe \xrightarrow[H_2SO_4]{NaN_3} PhCH{=}CHCONHMe$$

(91) (92)

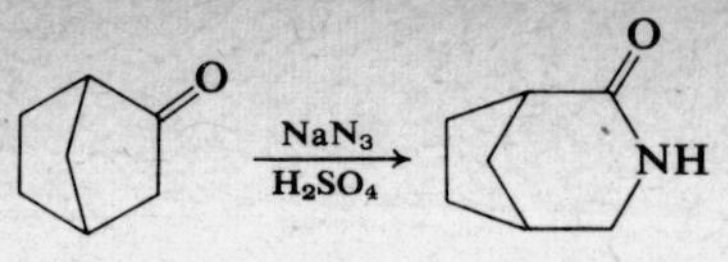

NaN_3 / H_2SO_4

(93) → (94)

Scheme 4

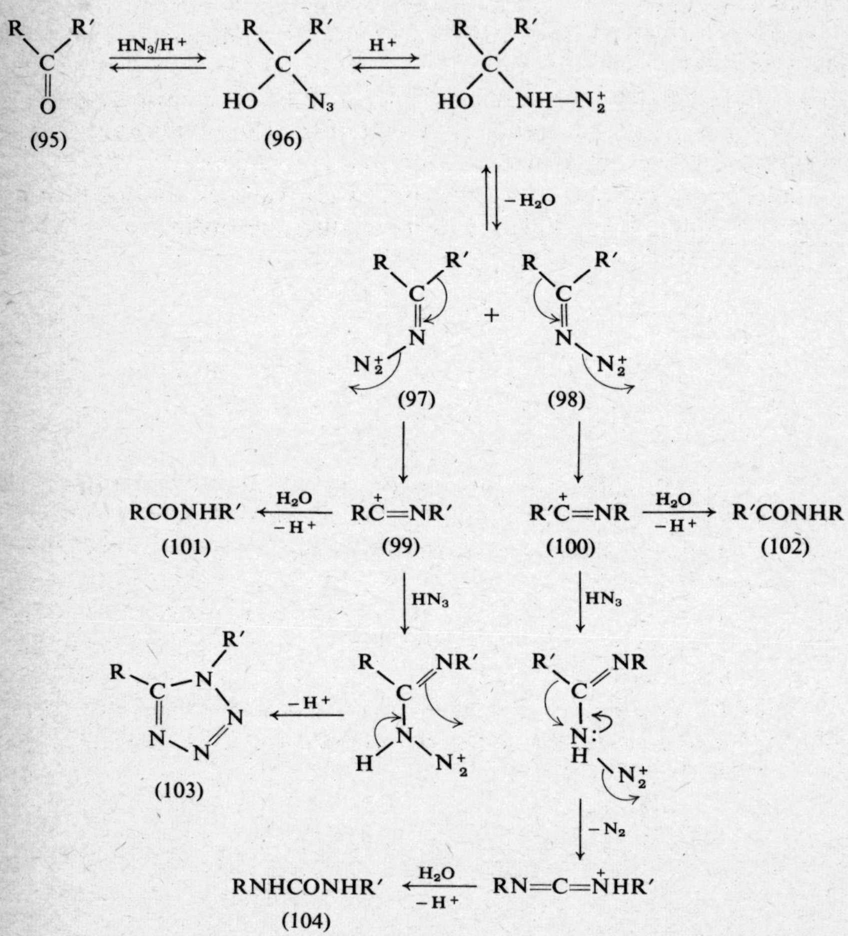

HN_3/H^+ ; H^+ ; $-H_2O$

(95) ⇌ (96) ⇌ HO, $NH—N_2^+$

(97) + (98)

RCONHR′ (101) ← H_2O, $-H^+$ — $R\overset{+}{C}{=}NR'$ (99)

$R'\overset{+}{C}{=}NR$ (100) — H_2O, $-H^+$ → R′CONHR (102)

HN_3

(103) ← $-H^+$

$-N_2$

RNHCONHR′ (104) ← H_2O, $-H^+$ — $RN{=}C{=}\overset{+}{N}HR'$

reaction is given in Scheme 4 (cf. 49 → 50; X = N_2^+) in which the first stage involves acid-catalysed addition of hydrogen azide to the carbonyl group. The key step in the reaction sequence (i.e., 97 → 99; 98 → 100) is closely analogous to that in the Beckmann rearrangement (*vide supra*). In reactions where unsymmetrical ketones (95; R $\neq$ R′) are used as substrates in the Schmidt rearrangement, it is not surprising, therefore, that the relative proportions of the isomeric amides (101 and 102) produced are closely similar to those found in the Beckmann rearrangement of the corresponding oximes when the latter reaction is carried out under equilibrating conditions (i.e., 60 $\rightleftharpoons$ 61). For example, treatment of the cyclopentanones (105) with hydrogen azide in sulphuric acid gives mixtures of the cyclic amides (106 and 107) in the same relative proportions as those formed by the Beckmann rearrangement of the corresponding oximes (108) in sulphuric acid [29].

R, =O, HN_3, H_2SO_4, R, NH, O, +, R, =O, NH, H_2SO_4, R, =NOH

(105) (106) (107) (108)

Evidence for the formation of the cationic intermediates (99 and 100) in the Schmidt reaction stems from the observation that tetrazoles (103) and substituted ureas (104) are often isolated when the reaction is carried out in the presence of an excess of azide [30]. These products must derive from intermediate species since amides themselves do not react with hydrogen azide under the reaction conditions.

As in the Beckmann rearrangement, various side-reactions are often observed in competition with Schmidt rearrangement. Fragmentation is commonly encountered in situations where a relatively stable carbonium ion can be eliminated from the diazonium-ion intermediates (97 and 98). For example, treatment of pinacolone (109) with hydrogen azide and sulphuric acid in the presence of benzene gives acetonitrile (110) and the t-butylbenzenes (111 and 112), formed by alkyla-

$MeCOCMe_3$ (109) $\xrightarrow[H_2SO_4]{HN_3}$ Me_3C, Me, C, N, N_2^+ $\xrightarrow{-N_2}$ $MeC{\equiv}N + Me_3C^+$ (110)

$\downarrow C_6H_6$

Me_3C—C₆H₄—CMe_3 + C₆H₅—CMe_3

(111) (112)

tion of benzene by the eliminated t-butyl cation [31]. Similarly, application of the reaction to aldehydes always results in the formation of nitriles (by proton elimination), although small amounts of formamides may also be formed, particularly in highly acidic media [32].

Neber reaction. The conversion of oxime tosylates (114; X = OTs), bearing an acidic hydrogen atom adjacent to the C=N function, to α-amino-ketones (117) or related compounds (118, 119) on treatment with strong base was discovered by Neber [33]. Related transformations of *N*,*N*-dichloroamines (113; X = Cl) have also been observed [34]. These reactions, which are thought [35] to proceed via an azirine intermediate (116) formed by intramolecular cyclisation of the conjugate base (115), are discussed more fully in Chapter 6 (pp. 138, 139).

$ArCH_2(R)CHNX_2$ (113) $\xrightarrow{-HX}$ $ArCH_2(R)C{=}N{-}X$ (114) $\xrightarrow{-H^+}$ $Ar\bar{C}H(R)C{=}N{-}X$ (115) $\xrightarrow{-X^-}$ azirine $Ar{-}CH$, $C{=}N$, R (116)

(116) $\xrightarrow{R'OH}$ $ArCHNH_2{-}RC(OR')_2$ (119)

(116) $\xrightarrow{H_2O}$ $ArCHNH_2{-}RC{=}O$ (117) $\xrightarrow[-2H]{-H_2O}$ pyrazine with Ar, R, N, Ar, R, N substituents (118)

Rearrangements of *N*-Substituted Amides

A number of organic reactions are known which may be described by the general mechanism outlined in Scheme 5. These rearrangements involve nucleophilic migration of a group R from a carbonyl carbon atom to an adjacent nitrogen atom that bears a group X that is stable as an anion or a neutral molecule. As in the case of the azide rearrangements discussed in a previous section, a concerted (120 → 122) or a two-step process (120 → 121 → 122) may be envisaged, the latter pathway proceeding via an acyl nitrene intermediate (121). In some examples the rearrangement product, an isocyanate (122), may actually be isolable. More commonly, further reaction occurs in the medium in which the rearrangement is conducted, and a variety of products may be obtained as indicated in Scheme 5.

Scheme 5

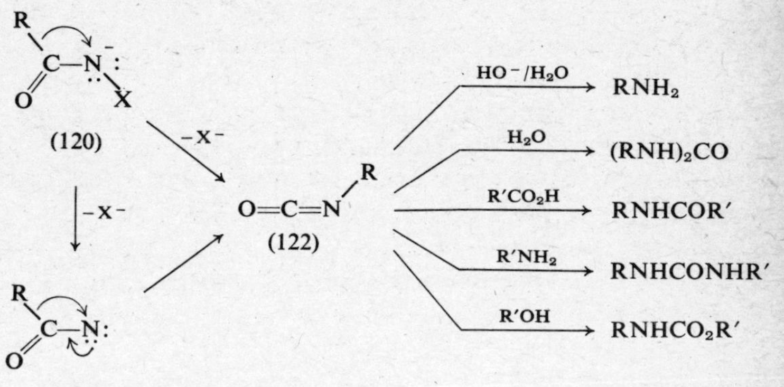

(120) $\xrightarrow{-X^-}$ $O{=}C{=}N{-}R$ (122)

(120) $\xrightarrow{-X^-}$ (121) $\rightarrow$ $O{=}C{=}N{-}R$ (122)

(122) $\xrightarrow{HO^-/H_2O}$ RNH_2

(122) $\xrightarrow{H_2O}$ $(RNH)_2CO$

(122) $\xrightarrow{R'CO_2H}$ $RNHCOR'$

(122) $\xrightarrow{R'NH_2}$ $RNHCONHR'$

(122) $\xrightarrow{R'OH}$ $RNHCO_2R'$

Hofmann rearrangement. The conversion of amides ($RCONH_2$) to amines (RNH_2) by treatment with bromine in sodium hydroxide solution was first reported [36] in 1881 by Hofmann, and constitutes one of the best known and most widely used reactions in organic chemistry. The reaction has found particular application in degradation sequences in which a carbon chain is shortened by one carbon atom. The intermediate isocyanate is not isolated. In alkaline solution, nucleophilic addition of hydroxide anion to the carbonyl group occurs, and spontaneous loss of carbon dioxide from the resulting carbamate anion ($RNHCO_2^-$) ensues. If the reaction is carried out in alcoholic solution, however, using alkoxide ($R'O^-$) as base in place of hydroxide, the resulting carbamates ($RNHCO_2R'$) can be isolated in good yield.

There is strong evidence that the Hofmann rearrangement proceeds via the conjugate base (120; X = Br) of the *N*-bromo-amide. Compounds of this type (RCONHX; X = Cl or Br) have been prepared independently and can be converted to isolable sodium salts ($RCO\bar{N}X\ Na^+$) which decompose spontaneously with rearrangement in protic solvents [37]. The Hofmann rearrangement may also be applied to the synthesis of amino-acids as shown by the conversion of succinimide (123) to *β*-alanine (124) on reaction with bromine in potassium hydroxide solution [38].

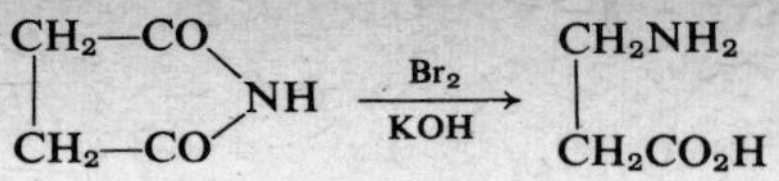

$$\begin{array}{l} CH_2\text{—}CO \\ \;|\qquad\qquad\; \rangle NH \\ CH_2\text{—}CO \end{array} \xrightarrow[KOH]{Br_2} \begin{array}{l} CH_2NH_2 \\ \;| \\ CH_2CO_2H \end{array}$$

(123) (124)

Lossen rearrangement. The conversion of hydroxamic acids (RCONHOH) or their *O*-acyl derivatives (RCONHOCOR′) to isocyanates on heating was discovered by Lossen in 1872 [39]. The rearrangement proceeds more readily in the presence of base to yield products derived from an intermediate isocyanate, by a mechanism closely akin to that of the Hofmann rearrangement (i.e. Scheme 5; X = OH or OCOR′). The Lossen rearrangement is less commonly encountered than the related Curtius and Hofmann procedures. However, a one-stage conversion of aromatic acids to amines ($ArCO_2H \rightarrow ArNH_2$) is made possible by an acid-catalysed variation of the reaction. The corresponding hydroxamic acids (ArCONHOH) are generated when the carboxylic acids are heated with nitromethane in polyphosphoric acid (which react to form hydroxylamine) and rearrange *in situ* to the protonated amines [40]. The use of the Lossen rearrangement in organic synthesis is illustrated [41] by the conversion of the ester (125) to epicamphor (126).

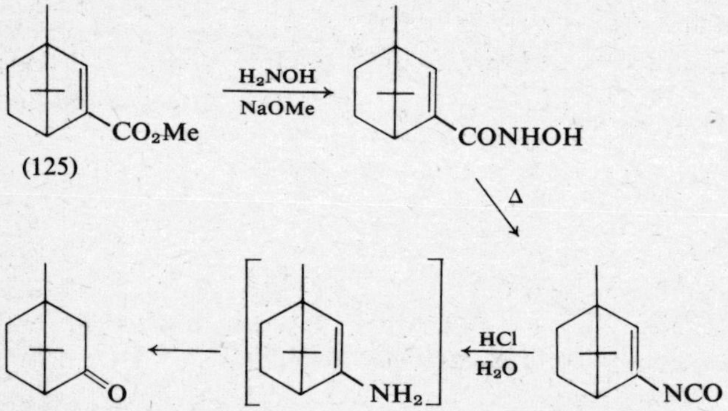

(125) CO_2Me $\xrightarrow[NaOMe]{H_2NOH}$ CONHOH $\xrightarrow{\Delta}$ NCO $\xrightarrow[H_2O]{HCl}$ [NH_2] $\rightarrow$ O (126)

Curtius rearrangement. The thermal rearrangement of acyl azides (127; i.e., 120; X = $N_2{}^+$) to isocyanates (122) was discovered [42] by Curtius in 1894, and it gives a synthetic route for the conversion of carboxylic acids to amines containing one carbon atom fewer. The azides may be prepared from the reaction of the acid chloride with sodium azide or by diazotisation of the acid hydrazide as shown in the following reaction sequence:

$$\begin{array}{l} \text{RCOCl} \xrightarrow{\text{NaN}_3} \\ \text{RCONHNH}_2 \xrightarrow{\text{HNO}_2} \end{array} \begin{array}{c} \text{RCON}_3 \\ (127) \end{array} \xrightarrow{-\text{N}_2} \begin{array}{c} \text{RNCO} \\ (122) \end{array}$$

The isocyanates themselves are isolated if the reaction is conducted in an inert aprotic solvent (e.g. benzene). More commonly, the rearrangement is carried out in solvents which react with isocyanates (e.g., $NaOH/H_2O$; $R'OH$; $R'NH_2$) and the corresponding amines, carbamates, or ureas are formed (see Scheme 5).

The mechanism of the Curtius rearrangement has been thoroughly investigated. The rate of the rearrangement of benzoyl azide (i.e., 127 $\rightarrow$ 122; R = Ph) was found to be first-order in azide and quite sensitive to the nature of the solvent medium, the fastest rates occurring in the more polar solvents [43]. Since the reaction rate is relatively insensitive to the presence and electronic character of substituents attached to the migrating benzene ring, neighbouring-group participation by this ring in the transition state for elimination of nitrogen is unimportant [44]. The formation of an acyl nitrene intermediate in these thermal reactions thus appears probable (i.e., 120 $\rightarrow$ 121 $\rightarrow$ 122; X = N_2^+). However, attempts to trap such an intermediate with radical scavengers (e.g. triphenylmethyl radical) have proved inconclusive and no initiation of polymerisation occurs when acyl azides are decomposed in the presence of vinyl monomers such as styrene. If a nitrene intermediate is formed in the Curtius rearrangement, therefore, conversion to isocyanate (121 $\rightarrow$ 122) must occur so fast that other reaction pathways are excluded.

The behaviour of sulphonyl azides (RSO_2N_3) provides an interesting contrast with that of their acyl counterparts. Benzenesulphonyl azide (128) does not rearrange on heating in benzene but decomposes with elimination of nitrogen and formation of the sulphonamides (130 and 131). The production of these compounds through reaction of an intermediate sulphonyl nitrene (129) with solvent molecules can readily be envisaged [45]. Curtius-type rearrangement of the azide (128) can be initiated photochemically, however. Ultraviolet irradiation of the compound in methanol solution leads to the formation of the sulphamate (132) in addition to the product (133) of capture of the nitrene intermediate by the solvent [46].

General mechanistic features. For the individual rearrangements discussed in this section, two reaction pathways can be considered by which the substrate rearranges either with concerted elimination of the

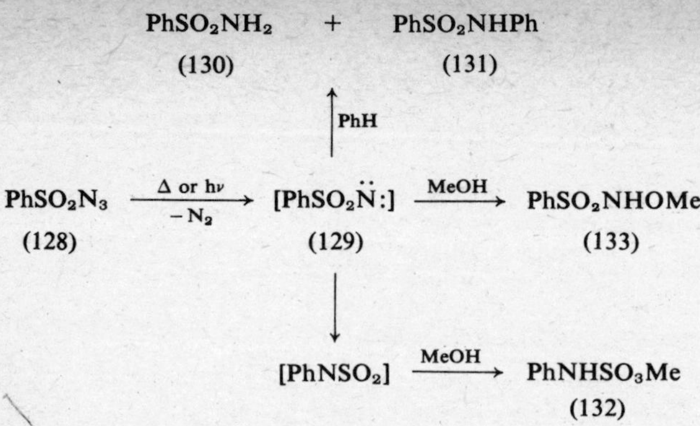

leaving-group (i.e., 120 → 122) or via a nitrenoid intermediate (121). The available evidence leads to the conclusion that the latter pathway is probable for thermally induced rearrangements. For reactions which are catalysed by acids or bases, however, the sensitivity of the rates of these processes to the nature of the migrating-group provides evidence, albeit tenuous, for a mechanism in which elimination of the leaving-group is assisted by participation of the migrating-group, i.e., a concerted rearrangement/elimination.

There is good evidence that, under most reaction conditions, the migrating-group never becomes fully detached from the substrate during the 1,2-shift. It has rigorously been established, for example, that the conversion of optically active 2-phenylpropionic acid ($PhCHMeCO_2H$) to 1-phenylethylamine ($PhCHMeNH_2$) using the Hofmann, Lossen, Curtius, or Schmidt procedures occurs with better then 95 per cent retention of configuration at the asymmetric carbon atom which migrates [21, 47]. In the rearrangement step, therefore, the new bond to the migration terminus (N) must begin to develop before the migrating-group has severed its connection with the migration origin (C).

References

1. T. Curtius and S. Darapsky, *J. prakt. Chem.*, 1901, **63**, 428.
2. C. L. Arcus *et al.*, *J. Chem. Soc.*, 1953, 178; 1954, 4319; 1957, 4064.
3. J. K. Senior, *J. Amer. Chem. Soc.*, 1916, **38**, 2718.
4. K. W. Sherk, A. G. Houpt, and A. W. Browne, *J. Amer. Chem. Soc.*, 1940, **62**, 329.
5. K. Biemann, G. Büchi, and B. H. Walker, *J. Amer. Chem. Soc.*, 1957, **79**, 5558.
6. J. H. Boyer and F. C. Canter, *J. Amer. Chem. Soc.*, 1955, **77**, 3287.

7. W. E. McEwen, M. Gilliland, and B. I. Sparr, *J. Amer. Chem. Soc.*, 1950, **72**, 3212.
8. W. H. Saunders and J. C. Ware, *J. Amer. Chem. Soc.*, 1958, **80**, 3328.
9. C. H. Gudmundsen and W. E. McEwan, *J. Amer. Chem. Soc.*, 1957, **79**, 329.
10. C. L. Arcus and E. A. Lucken, *J. Chem. Soc.*, 1955, 1634.
11. Y. Yukawa and K. Tanaka, see *Chem. Abs.*, 1958, **52**, 2795, 4513.
12. J. Stieglitz and P. N. Leech, *J. Amer. Chem. Soc.*, 1914, **36**, 272.
13. M. Appl and R. Huisgen, *Chem. Ber.*, 1959, **92**, 2961.
14. E. Beckmann, *Ber.*, 1886, **19**, 988.
15. W. E. Bachmann and M. X. Barton, *J. Org. Chem.*, 1938, **3**, 300.
16. C. S. Marvel and J. C. Eck, *Org. Synth. Coll. Vol.* **2**, 1943, 71.
17. H. N. Rydon, N. H. P. Smith, and D. Williams, *J. Chem. Soc.*, 1957, 1900.
18. M. Martynoff, *Ann. Chim.*, 1937, **7**, 424.
19. M. Kuhara, Y. Todo, and H. Watanabe, see *Chem. Abs.*, 1911, **5**, 1278; 1917, **11**, 579.
20. J. Meisenheimer, P. Zimmermann, and U. von Kummer, *Annalen*, 1926, **446**, 205.
21. J. Kenyon and D. P. Young, *J. Chem. Soc.*, 1941, 263.
22. R. T. Conley, *J. Org. Chem.*, 1963, **28**, 278.
23. R. F. Brown, N. M. van Gulick, and G. H. Schmidt, *J. Amer. Chem. Soc.*, 1955, **77**, 1094.
24. A. Werner and A. Piguet, *Ber.*, 1904, **37**, 4295.
25. K. F. Schmidt, *Z. angew. Chem.*, 1923, **36**, 511.
26. A. I. Vogel, *A Textbook of Practical Organic Chemistry*, Third edn., p. 919; Longmans (London), 1956.
27. L. H. Briggs, G. C. DeAth, and S. R. Ellis, *J. Chem. Soc.*, 1942, 61.
28. R. C. Elderfield and E. T. Losin, *J. Org. Chem.*, 1961, **26**, 1703.
29. H. Schechter and J. C. Kirk, *J. Amer. Chem. Soc.*, 1951, **73**, 3087.
30. P. A. S. Smith, *J. Amer. Chem. Soc.*, 1948, **70**, 320.
31. H. D. Zook and S. C. Paviak, *J. Amer. Chem. Soc.*, 1955, **77**, 2501.
32. W. E. McEwan, W. E. Conrad, and C. A. VanderWerf, *J. Amer. Chem. Soc.*, 1952, **74**, 1168.
33. P. W. Neber, A. Burgard, and W. Thier, *Annalen*, 1932, **493**, 281; 1936, **526**, 277.
34. H. E. Baumgarten, F. A. Bower, and J. M. Petersen, *J. Amer. Chem. Soc.*, 1954, **76**, 4561; 1960, **82**, 459.
35. D. J. Cram and M. J. Hatch, *J. Amer. Chem. Soc.*, 1953, **75**, 33, 38.
36. A. W. Hofmann, *Ber.*, 1881, **14**, 2725.
37. C. Mauguin, *Ann. Chim.*, 1911, **22**, 297.
38. H. T. Clarke and L. D. Behr, *Org. Synth. Coll. Vol.* **2**, 1943, 19.
39. H. Lossen, *Annalen*, 1872, **161**, 347.
40. G. B. Bachman and J. E. Goldmacher, *J. Org. Chem.*, 1964, **29**, 2576.
41. J. Bredt and W. H. Perkin, *J. Chem. Soc.*, 1913, 2182.

42. T. Curtius, *Ber.*, 1894, **27**, 778.
43. M. S. Newman, S. H. Lee, and A. B. Garrett, *J. Amer. Chem. Soc.*, 1947, **69**, 113.
44. Y. Yukawa and Y. Tsuno, *J. Amer. Chem. Soc.*, 1957, **79**, 5530.
45. O. C. Dermer and M. T. Edmison, *J. Amer. Chem. Soc.*, 1955, **77**, 70.
46. W. Lwowski and E. Scheiffele, *J. Amer. Chem. Soc.*, 1965, **87**, 4359.
47. J. Kenyon *et al.*, *J. Chem. Soc.*, 1935, 1072; 1939, 916; 1946, 25.

Review Articles

General reviews

P. A. S. Smith in *Molecular Rearrangements* Part 1; pp. 462–568; ed. de Mayo, Interscience (New York), 1963.

N. V. Sidgwick, *The Organic Chemistry of Nitrogen*, 3rd edn., revised by I. T. Millar and H. D. Springall, Clarendon Press (Oxford), 1966.

D. V. Banthorpe in *The Chemistry of the Amino Group*, pp. 585–668; ed. Patai, Interscience (New York), 1968.

C. G. McCarty in *The Chemistry of the C–N Double Bond*, pp. 363–464; ed. Patai, Interscience (New York), 1970.

D. V. Banthorpe in *The Chemistry of the Azido Group*, pp. 397–440; ed. Patai, Interscience (New York), 1971.

Schmidt reactions

H. Wolff, *Org. Reactions*, 1946, **3**, 307.

Beckmann rearrangement

F. D. Popp and W. E. McEwen, *Chem. Rev.*, 1958, **58**, 370.

L. G. Donaruma and W. Z. Heldt, *Org. Reactions*, 1960, **11**, 1.

F. Uhlig and H. R. Snyder, *Adv. Org. Chem.*, 1960, **1**, 35.

M. I. Vinnik and N. G. Zarakhani, *Russian Chem. Rev.*, 1967, 51.

Neber reactions

C. O'Brien, *Chem. Rev.*, 1964, **64**, 81.

Curtius rearrangement

P. A. S. Smith, *Org. Reactions*, 1946, **3**, 337.

Lossen rearrangement

H. L. Yale, *Chem. Rev.*, 1943, **33**, 209.

4

Nucleophilic Rearrangements III: Migration from Carbon to Oxygen

In the preceding two chapters, nucleophilic migration of carbon groups to carbon and nitrogen atom termini has been discussed. Related rearrangements in which an oxygen atom is located at the migration terminus are also known and reactions of this type will be dealt with in this chapter. With few exceptions, such rearrangements are restricted to compounds containing a peroxy grouping (O—O—) and there is consequently much less diversity in reaction type than that previously encountered for the C-to-C and C-to-N rearrangements.

Scheme 1

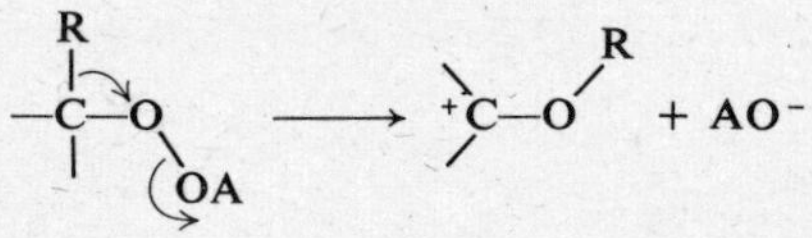

$$\text{—}\underset{|}{\overset{\text{R}}{\overset{|}{\text{C}}}}\text{—O—OA} \longrightarrow {}^{+}\text{C—O—R} + \text{AO}^{-}$$

A general mechanism for these reactions can be written as shown in Scheme 1. As in the case of the other nucleophilic rearrangements discussed in previous chapters, two events occur during the course of the reaction, namely, elimination of a leaving-group (in this case OA) with its pair of bonding electrons, and nucleophilic shift of the migrating-group. We have learned previously that these two events may occur either in a stepwise or in a synchronous fashion. For the C-to-O rearrangements, however, departure of the leaving-group OA prior to shift of the migrating group R would create at the migration terminus a univalent positively charged oxygen atom with six electrons only in its valence shell (i.e. $R_3COOA \rightarrow R_3CO^+ + AO^-$). Energetically, such a process is highly unfavourable and much less probable than, for example, analogous reactions leading to the production of trivalent C^+ or divalent N^+ centres. The operation of a synchronous process (Scheme 1), of course, prevents the formation of an R_3CO^+ intermediate and all of the reactions described in this chapter follow such a pathway.

Rearrangements of Peroxy-esters

The spontaneous rearrangement of peroxy-esters of the type (1) to give oxygen-bridged bicycloalkyl esters (2) was first described [1] in 1944 and the reaction is sometimes referred to as the **Criegee rearrangement**. The products, in fact, are esters of internal hemiketals and, in consequence, undergo ready hydrolysis to the corresponding hydroxycycloalkanones (3). This reaction sequence thus provides a convenient synthetic route to compounds of the cyclononane and cyclodecane series from the readily accessible hydrindane and decalin precursors (1; n = 3 and 4 respectively).

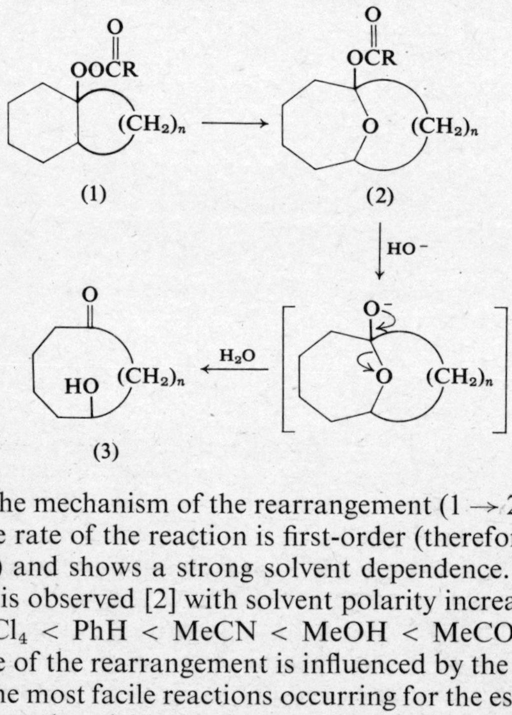

Studies of the mechanism of the rearrangement (1 → 2) have established that the rate of the reaction is first-order (therefore a unimolecular process) and shows a strong solvent dependence. Thus, an increase in rate is observed [2] with solvent polarity increasing through the series: CCl_4 < PhH < MeCN < MeOH < $MeCO_2H$. Furthermore, the ease of the rearrangement is influenced by the nature of the ester group, the most facile reactions occurring for the esters of strong acids. For example, trichloroacetates rearrange faster than acetates (1; R = Cl_3C and Me respectively) and *p*-nitrobenzoates faster than benzoates (1; R = *p*-$O_2NC_6H_4$ and Ph respectively). These results are in harmony with a mechanism in which the rate-determining step involves heterolytic fission of the weak O—O bond and liberation of a carboxylate anion (RCO_2^-). The rate of such a process would obviously be sensitive to the ionising power of the solvent and would be accelerated in cases where the liberated anion is stabilised by electron-withdrawing properties in the group R.

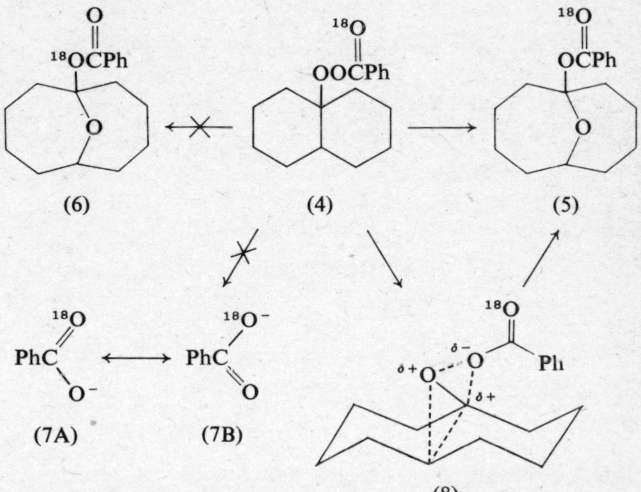

(6) (4) (5)

(7A) (7B) (8)

Dissociation of the peroxy-ester into *free* ions, however, cannot occur during the rearrangement. This became apparent when it was found [3] that the ester (2; R = Ph, n = 4) is the sole product of rearrangement of decalyl perbenzoate (1; R = Ph, n = 4) in the presence of p-substituted benzoate anions, no incorporation of these ions occurring. More intriguingly, it was also found [4] that rearrangement of the ester (4) containing an ^{18}O-labelled carbonyl group gives exclusively the product (5) in which the label is retained by the carbonyl group. If a *free* benzoate anion had been liberated during the rearrangement, the resulting equivalence of its two oxygen atoms (cf. 7A ↔ 7B) would have led to the additional formation of the differently labelled ester (6).

In order to accommodate these seemingly contradictory results, it has been proposed that the reaction proceeds by formation of a tight ion-pair which collapses to product before equilibration of the carboxylate oxygen atoms can occur (e.g., 4 → 8 → 5). Intermediates of this type (8) still retain a measure of covalent character (i.e., ion formation is not complete) and the rearrangement can be described as a synchronous movement of groups in which two new C—O bonds are forming as the O—O and C—C bonds are breaking (cf. broken lines in structure 8). Since the reaction exhibits the characteristics of a polar process, however, considerable separation of charge must develop during the rearrangement step.

Me_2C—O—OCOPh (9) ⟶ [$Me_2C^{\delta+}$—$O^{\delta+}$—$O^{\delta-}$COPh, δ+] ⟶ Me_2C—O(Ph)—OCOPh (10)

(11) ⟶ (12) OCOAr / OPh $\xrightarrow{H_2O}$ =O + PhOH + $ArCO_2H$

This rearrangement is not restricted to the peroxy-esters of bicycloalkanes. For example, α-cumyl perbenzoate (9) rearranges to the ester (10) with the exclusive migration of a phenyl group [5]. A similar preference for aryl over alkyl migration is evident in the rearrangement of the *p*-nitroperbenzoate (11; Ar = *p*-$O_2NC_6H_4$) which yields products derived from the hemiketal ester (12) by hydrolysis [6]. In reactions where different groups may potentially migrate to oxygen, therefore, preferential migration of the group most capable of sustaining positive charge is observed.

Rearrangements of Hydroperoxides

Hydroperoxides (R_3COOH) do not rearrange spontaneously like their ester derivatives since the hydroxyl anion is not a sufficiently good leaving-group to initiate reaction. Very ready reactions do occur, however, in the presence of protic or Lewis acids. The conversion of t-alkyl hydroperoxides to a mixture of ketone and alcohol or phenol (i.e., $R_3COOH \rightarrow R_2CO + ROH$) in the presence of acid was first reported [7] by Wieland, although the scope of the reaction was not properly explored until several years later [8]. Representative examples of this reaction are provided by the following transformations, the second of which proceeds by ring opening to give a hydroxyketone [9].

$$Ph_3COOH \xrightarrow{H^+} Ph_2C{=}O + PhOH$$

$$\text{1-methylcyclohexyl hydroperoxide (Me, OOH)} \xrightarrow{H^+} \text{COMe}(CH_2)_4CH_2OH$$

Hydroperoxides are usually synthesised by the reaction of alcohols with hydrogen peroxide ($ROH + H_2O_2 \rightarrow ROOH + H_2O$). If this preparation is carried out in strongly acidic media, rearrangement of the hydroperoxide *in situ* is brought about. Similarly, peroxidation of alkenes in the presence of a strong acid leads to the formation of products derived from an intermediate hydroperoxide. These reactions, exemplified below, avoid the necessity of isolation of the hydroperoxide, and may be likened to the Schmidt reactions of alcohols and alkenes with hydrogen azide (see Chapter 3).

$$Ar_3COH \xrightarrow[HClO_4]{H_2O_2} Ar_2C{=}O + ArOH$$

$$Ph_2C{=}CH_2 \xrightarrow[H_2SO_4]{H_2O_2} PhCOMe + PhOH$$

The hydroperoxide rearrangement has important commercial application, for example in the industrial production of phenol and acetone from benzene. Cumene ($PhCHMe_2$), which is cheaply available from the Friedel–Crafts alkylation of benzene with propene, readily undergoes autoxidation to α-cumyl hydroperoxide ($PhCMe_2OOH$) and thence to phenol and acetone by acid-promoted rearrangement.

The mechanism of the hydroperoxide rearrangement has been subjected to thorough investigation by a number of workers, notably Kharasch [10]. It has been established that overall the reaction rate is second-order, first-order in both hydroperoxide and acid concentration [i.e. Rate $= k(R_3COOH)(\text{acid})$]. The appearance of acid concentration in the rate expression suggests that the rearranging species is a conjugate acid of the hydroperoxide. In acidic solution, equilibria between the hydroperoxide (13) and its conjugate acids (14 and 15) will be established. The rearrangement most probably proceeds via the latter intermediate (15) by concerted elimination of a water molecule and shift of the migrating-group as indicated in Scheme 2.

Scheme 2

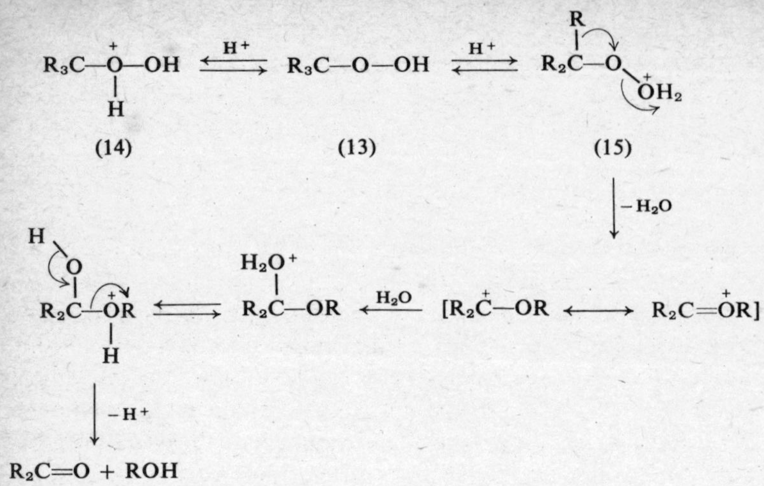

$$\underset{(14)}{R_3C\text{—}\overset{+}{O}(H)\text{—}OH} \xrightleftharpoons{H^+} \underset{(13)}{R_3C\text{—}O\text{—}OH} \xrightleftharpoons{H^+} \underset{(15)}{R_2C(R)\text{—}O\text{—}\overset{+}{O}H_2}$$

$$\downarrow -H_2O$$

$$R_2C(OH)\text{—}\overset{+}{O}R(H) \rightleftharpoons R_2C(H_2O^+)\text{—}OR \xleftarrow{H_2O} [R_2\overset{+}{C}\text{—}OR \longleftrightarrow R_2C{=}\overset{+}{O}R]$$

$$\downarrow -H^+$$

$$R_2C{=}O + ROH$$

In agreement with this mechanism, it has been found that the ease of the rearrangement is proportional to the ability of the migrating-group to share positive charge. Thus, aryl groups migrate in preference to alkyl groups and, for a series of alkyl groups, the migratory aptitude increases through the series: primary (RCH_2) < secondary (R_2CH) < tertiary (R_3C). In the case of aryl group shifts, electron-releasing substituents attached to the benzene ring increase the tendency to migrate and electron-attracting substituents have the opposite effect. A similar pattern of behaviour to that established for other nucleophilic rearrangements is therefore evident.

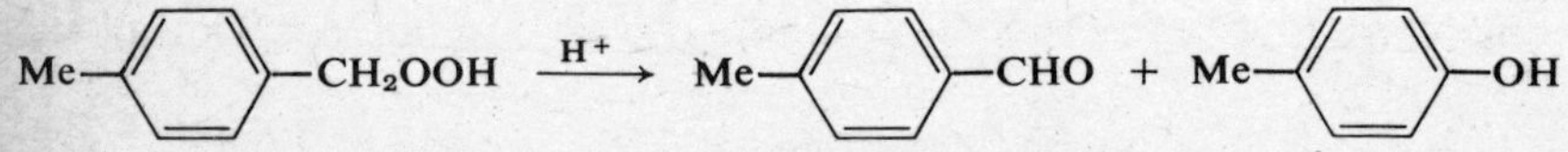

$$\underset{(16)}{Me\text{—}C_6H_4\text{—}CH_2OOH} \xrightarrow{H^+} \underset{(17)}{Me\text{—}C_6H_4\text{—}CHO} + \underset{(18)}{Me\text{—}C_6H_4\text{—}OH}$$

$$\underset{(19)}{\text{cyclohex-2-enyl}(H)(OOH)} \xrightarrow{H^+} \underset{(20)}{OHC\text{—}(CH_2)_4\text{—}CHO} + \underset{(21)}{\text{cyclopent-1-enyl}\text{—}CHO}$$

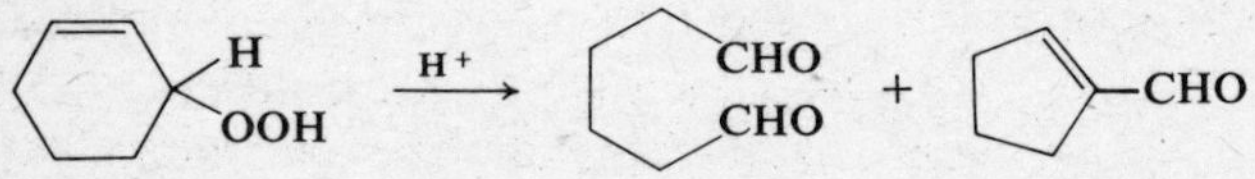

For rearrangements of benzylic hydroperoxides, hydrogen migration is usually found to occur in competition with aryl migration.

Thus, acid-promoted decomposition of *p*-methylbenzyl hydroperoxide (16) gives a mixture of *p*-tolualdehyde (17) (hydrogen migration) and *p*-cresol (18) (aryl migration) in the ratio of about 1·6 to 1 respectively. Vinyl groups have been found to migrate in preference to alkyl groups. For example, the rearrangement of the cyclohexenyl hydroperoxide (19) gives a mixture of adipaldehyde (20) and the formylcyclopentene (21), the product of self-condensation of adipaldehyde under the reaction conditions.

Baeyer–Villiger Reaction

The oxidation of ketones to esters with Caro's acid (i.e. peroxysulphuric acid; H_2SO_5) was discovered [11] in 1899 by von Baeyer and Villiger. Later, it was found that the reaction can also be effected by organic peracids (e.g. perbenzoic acid; $PhCO_3H$) as well as by hydrogen peroxide itself in weakly basic solution. Since the reaction is applicable both to aryl and alkyl ketones, it has proved most useful in synthesis and particularly for the selective degradation of complex molecules. Representative examples of the Baeyer–Villiger reaction are given below.

$$PhCOPh \xrightarrow{RCO_3H} PhCO_2Ph$$

$$MeCOPh \xrightarrow{RCO_3H} MeCO_2Ph$$

Cyclohexanone (=O) $\xrightarrow{RCO_3H}$ [seven-membered lactone, O, =O] $\xrightarrow{EtOH}$ CH_2OH … CO_2Et

The mechanism of the reaction has been studied by several groups. Doering showed [12] that perbenzoic acid oxidation of benzophenone labelled with ^{18}O gives phenyl benzoate in which the ^{18}O-label is retained by the carbonyl group. The intramolecular nature of the migration has also been established, and it has been shown that the migrating-group shifts with complete retention of configuration, e.g., in the oxidation of optically active 1-phenylethyl methyl ketone to the corresponding acetate (i.e. PhCHMeCOMe ⟶ PhCHMeOCOMe) [13]. Studies of the rate of the acid-promoted reaction have established [14] a third-order rate expression [i.e., Rate = k(ketone)(peracid)(acid)] in harmony with a mechanism in which the transition state in the rate-determining step of the process is derived from all three reagents.

Scheme 3

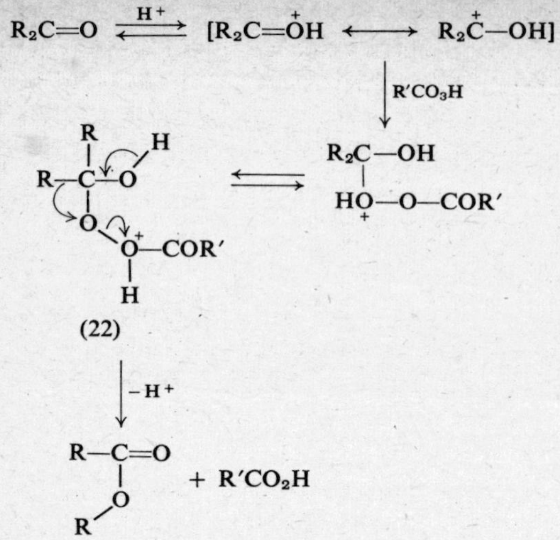

$$R_2C{=}O \xrightleftharpoons{H^+} [R_2C{=}\overset{+}{O}H \longleftrightarrow R_2\overset{+}{C}{-}OH]$$

$$\downarrow R'CO_3H$$

$$R_2C(OH){-}\overset{+}{O}H{-}O{-}COR' \rightleftharpoons R_2C(OH){-}O{-}\overset{+}{O}(H){-}COR' \quad (22)$$

$$\downarrow -H^+$$

$$R{-}C(=O){-}O{-}R + R'CO_2H$$

A mechanism which accounts nicely for all of these observations is given in Scheme 3. The key step involves the conjugate acid (22) and proceeds by nucleophilic migration of R to the oxygen atom from which the leaving-group ($R'CO_2H$) is eliminated. In basic solution, the oxidation of ketones by hydrogen peroxide probably occurs by rearrangement of the anion (23), the product of nucleophilic addition of hydroperoxide anion to the carbonyl group of the ketone.

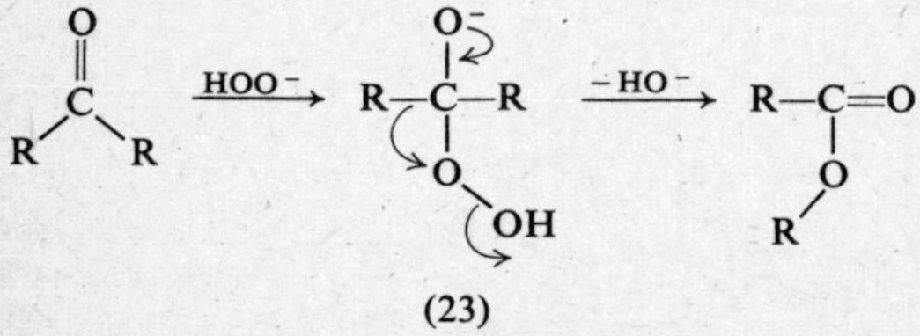

$$R_2C{=}O \xrightarrow{HOO^-} R_2C(O^-){-}O{-}OH \;(23) \xrightarrow{-HO^-} R{-}C(=O){-}O{-}R$$

$$(24) \xrightarrow{RCO_3H} (25) + (26)$$

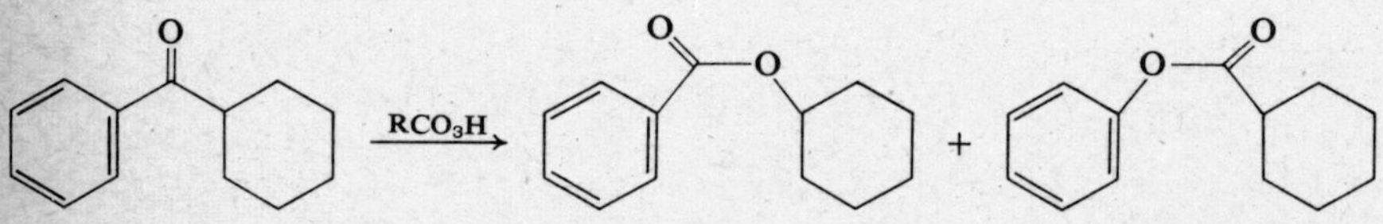

(24) (25) (26)

In principle, the Baeyer-Villiger oxidation of an unsymmetrical ketone can give rise to two isomeric esters and, in practice, two such products are often formed (i.e., $RCOR' \rightarrow RCO_2R' + R'CO_2R$).

For example, peracid oxidation of phenyl cyclohexyl ketone (24) yields cyclohexyl benzoate (25) and phenyl cyclohexanecarboxylate (26). The relative proportions of esters formed from an unsymmetrical ketone will clearly be related to the relative migratory aptitudes of the groups which may shift to oxygen in the rearrangement step (see Scheme 3). The effect of *p*-substitution on the tendency of phenyl groups to migrate has been studied by Doering using a series of substituted benzophenones (*p*-XC_6H_4COPh) [15]. It was found that an electron-releasing substituent increases the tendency to migrate while an electron-attracting substituent reduces the mobility of the group. Thus, the following order of preference for migration was found: *p*-$MeOC_6H_4$ > *p*-MeC_6H_4 > Ph > *p*-ClC_6H_4 > *p*-$O_2NC_6H_4$. This pattern of behaviour should now be familiar for reactions involving a 1,2-nucleophilic migration. For alkyl groups, the migratory aptitude increases through the series: Me < primary (RCH_2) < secondary (R_2CH) < tertiary (R_3C) [16].

It has also become apparent, however, that the relative migratory aptitudes of groups vary with the nature of the peracid oxidising agent and therefore with the nature of the leaving-group in the rearrangement (see Scheme 3). For example, oxidation of the ketone (24) gives the esters (25) and (26) in the relative proportions 9:1 respectively using peracetic acid, and 4:1 respectively using trifluoroperacetic acid. In both reactions, however, a preference for cyclohexyl over phenyl migration was found.

Migratory aptitudes are also extremely sensitive to steric and structural effects. This is strikingly illustrated by the behaviour of ketones possessing a bicyclo[2.2.1]heptanone skeleton. Oxidation of epicamphor (27) gives exclusively the lactone (28); migration of the secondary bridgehead carbon atom occurs in preference to shift of

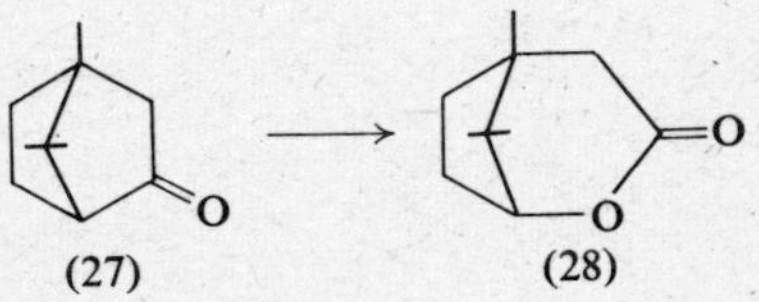

(27) → (28)

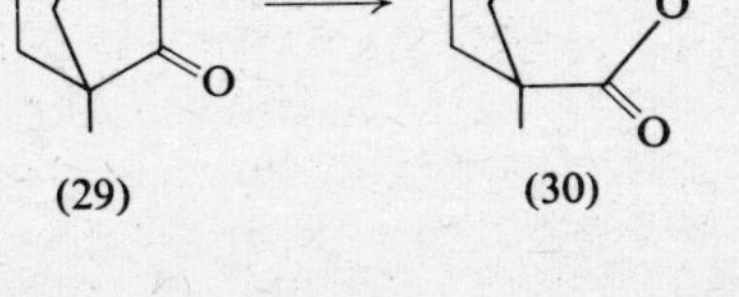

(29) → (30)

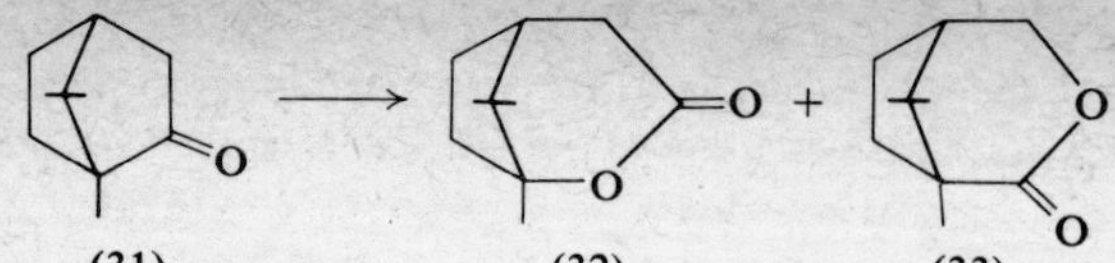

(31) (32) (33)

the methylene group. On the other hand, 1-methylnorcamphor (29) gives only the product (30) of methylene migration, and camphor (31) gives a mixture of lactones (32 and 33) where both possible rearrangements occur [17].

Dakin Reaction

The oxidation of *o*- and *p*-hydroxybenzaldehyde to catechol and quinol respectively on treatment with alkaline hydrogen peroxide was first reported by Dakin in 1909 [18].

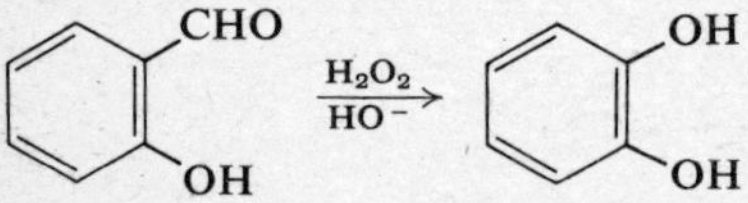

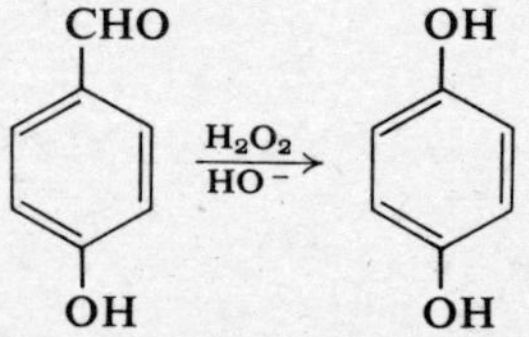

$$\text{CHO, OH (ortho)} \xrightarrow[\text{HO}^-]{\text{H}_2\text{O}_2} \text{OH, OH (ortho)} \qquad \text{CHO, OH (para)} \xrightarrow[\text{HO}^-]{\text{H}_2\text{O}_2} \text{OH, OH (para)}$$

The mechanism of the Dakin reaction probably is closely similar to that of the Baeyer–Villiger reaction in alkaline media (*vide supra*), as indicated in Scheme 4. Nucleophilic addition of hydroperoxide anion to the carbonyl group of the aldehyde produces an intermediate (34)

Scheme 4

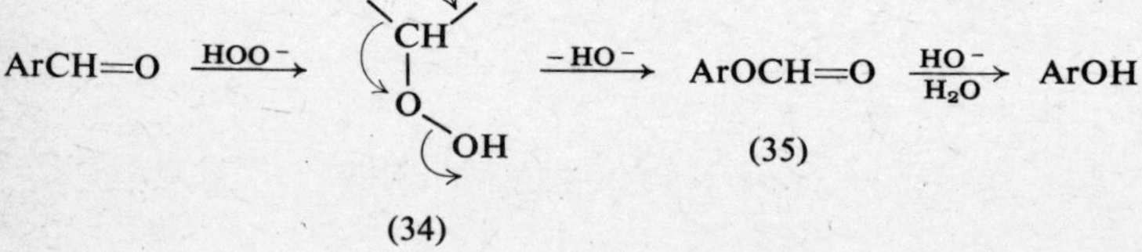

$$\text{ArCH{=}O} \xrightarrow{\text{HOO}^-} \text{Ar(O}^-\text{)CH{-}O{-}OH}\ (34) \xrightarrow{-\text{HO}^-} \text{ArOCH{=}O}\ (35) \xrightarrow[\text{H}_2\text{O}]{\text{HO}^-} \text{ArOH}$$

$$\text{(36)}\ o\text{-HO{-}C}_6\text{H}_4\text{{-}CH(O}^-\text{){-}O{-}OH} \longrightarrow o\text{-}^-\text{O{-}C}_6\text{H}_4\text{{-}OCH{=}O} + \text{H}_2\text{O}$$

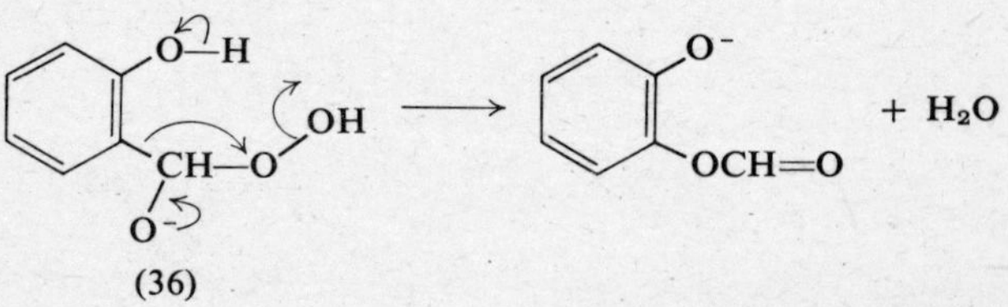

which can rearrange to give a formate (35). In the case of salicylaldehyde, the very ready reaction [19] probably proceeds via a cyclic transition state (36), the phenolic proton being 'picked up' by the hydroxide leaving-group. Under the alkaline conditions of the reaction, the formate (35) is hydrolysed to the corresponding phenol.

Benzaldehyde itself is oxidised to benzoic acid by alkaline hydrogen peroxide and only trace amounts of phenol are formed. Most substituted benzaldehydes show similar behaviour. In this case, migration of hydride in the intermediate (34) is strongly favoured over 1,2-shift of the benzene ring. The presence of an *o*- or *p*-hydroxy substituent, however, substantially increases the migratory aptitude of the aryl ring such that rearrangement takes place. An *o*- or *p*-amino substituent can provide the same assistance and the oxidation of aminobenzaldehydes to aminophenols occurs smoothly.

References

1. R. Criegee, *Ber.*, 1944, **77**, 722.
2. R. Criegee and R. Kaspar, *Annalen*, 1948, **560**, 127.
3. P. D. Bartlett and J. L. Kice, *J. Amer. Chem. Soc.*, 1953, **75**, 5591; H. L. Goering and A. C. Olsen, *J. Amer. Chem. Soc.*, 1953, **75**, 5853.
4. D. B. Denney and D. G. Denney, *J. Amer. Chem. Soc.*, 1957, **79**, 4806.
5. H. Hock and H. Kropf, *Chem. Ber.*, 1955, **88**, 1544.
6. H. Kwart and R. T. Keen, *J. Amer. Chem. Soc.*, 1959, **81**, 943.
7. H. Wieland and J. Maier, *Ber.*, 1931, **64**, 1205.
8. W. Dilthey, F. Quint, and H. Dierichs, *J. prakt. Chem.*, 1938, **151**, 25.
9. W. Pritzkow and K. A. Müller, *Chem. Ber.*, 1956, **89**, 2321.
10. M. S. Kharasch *et al.*, *J. Org. Chem.*, 1952, **17**, 207 and earlier papers.
11. A. von Baeyer and V. Villiger, *Ber.*, 1899, **32**, 3625.
12. W. von E. Doering and E. Dorfmann, *J. Amer. Chem. Soc.*, 1953, **75**, 5595.
13. K. Mislow and J. Bremner, *J. Amer. Chem. Soc.*, 1953, **75**, 2319.
14. M. F. Hawthorne and W. D. Emmons, *J. Amer. Chem. Soc.*, 1958, **80**, 6398.
15. W. von E. Doering and L. Speers, *J. Amer. Chem. Soc.*, 1950, **72**, 5515.
16. M. F. Hawthorne, W. D. Emmons, and K. S. McCallum, *J. Amer. Chem. Soc.*, 1958, **80**, 6393.
17. R. R. Sauers and G. P. Ahearn, *J. Amer. Chem. Soc.*, 1961, **83**, 2759.
18. H. D. Dakin, *Amer. Chem. J.*, 1909, **42**, 477.
19. A. Wacek, H. O. Eppinger, and A. V. Bezard, *Ber.*, 1940, **73**, 644; 1941, **74**, 845.

Review Articles

General reviews

P. A. S. Smith in *Molecular Rearrangements* Part 1; pp. 568–591; ed. de Mayo, Interscience (New York), 1963.

J. B. Lee and B. C. Uff, *Quart. Rev.*, 1967, **21**, 429.

Hydroperoxide rearrangements

J. E. Leffler, *Chem. Rev.*, 1949, **45**, 385.

P. D. Bartlett, *Rec. Chem. Progr.*, 1950, **11**, 47.

Baeyer–Villiger reaction

C. H. Hassall, *Org. Reactions*, 1957, **9**, 73.

H. O. House, *Modern Synthetic Reactions*, pp. 123–129, Benjamin (New York), 1965.

5

1, 2-Electrophilic Rearrangements

It is natural to envisage a type of molecular rearrangement, complementary to the nucleophilic migrations discussed in Chapters 2, 3, and 4, in which the migrating-group proceeds to an *electron-rich* destination—an *electrophilic* rearrangement; that is:

$$-\overset{..}{\underset{|}{A}}-\overset{\overset{R}{|}}{\underset{|}{B}}- \longrightarrow -\overset{\overset{R}{|}}{\underset{|}{A}}-\overset{..}{\underset{|}{B}}-$$

Rearrangements of this form have come to light in great variety, as illustrated in Table 1. This chapter is devoted to 1,2-migrations, with some closely allied extensions, which make a coherent narrative. Migrations over a longer range will be treated in the following chapter.

The first case to be studied in detail (Stevens, 1928) may be exemplified as follows:

$$\underset{(1)}{PhCOCH_2\overset{+}{N}Me_2(CH_2Ph)} \xrightarrow[\text{or } HO^-]{EtO^-} \underset{(2)}{PhCO\overset{-}{C}H\overset{+}{N}Me_2(CH_2Ph)} \longrightarrow \underset{(3)}{PhCOCH(CH_2Ph)NMe_2}$$

Here, the destination atom in (2) is strongly nucleophilic. This was regarded as providing the main driving force of the reaction, long considered to be an internal S_N2 displacement on the side-chain carbon atom of the migrating benzyl, ammonium nitrogen being a favourable leaving-group.

We shall first describe the behaviour of systems in which the destination atom is anionic carbon, then those with another destination atom. More recent work, in which the mechanism of the reaction undergoes drastic reconsideration, will then be discussed. The stereo-

Table 1. Rearrangements Involving 1,2-Migration

Type	Example			Reagent
$N^+ \rightarrow C^-$	$(PhCH_2)_2\overset{+}{N}Me_2$	$\rightarrow$	NMe_2 | $PhCHCH_2Ph$	$PhLi$ or H_2N^-
$O \rightarrow C^-$	$(PhCH_2)_2O$	$\rightarrow$	OH | $PhCHCH_2Ph$	$PhLi$ or H_2N^-
$N \rightarrow C^-$	$PhCOCH_2NPh$ | CH_2Ph	$\rightarrow$	$PhCOCHNHPh$ | CH_2Ph	KOH fusion
$C \rightarrow C^-$	Ph_3CCH_2Cl	$\rightarrow$	Ph_2CHCH_2Ph	Na
$N^+ \rightarrow O^-$	$PhCH_2\overset{+}{N}Me_2$ | O^-	$\rightarrow$	$PhCH_2ONMe_2$	Heat
$N^+ \rightarrow N^-$	$MeCO\overset{-}{N}\overset{+}{N}Me_2$ | CH_2Ar	$\rightarrow$	$MeCONNMe_2$ | CH_2Ar	Heat
$S^+ \rightarrow C^{-}$ [a]	$PhCOCH{=}SMe$ | CH_2Ph	$\rightarrow$	$PhCOCHCH_2Ph$ | SMe	H_2O (100°)
$S^+ \rightarrow N^-$	$Ts\overset{-}{N}\overset{+}{S}CH_2Ph$ | C_3H_5	$\rightarrow$	$TsNSCH_2Ph$ | C_3H_5	Heat

[a] ($\overset{+}{S}{-}\overset{-}{C} \leftrightarrow S{=}C$) (the sulphur octet can expand to a decet).

chemistry of the rearrangements, the behaviour of allylic systems, and the notable phenomena observed when the migration origin is a sulphur atom, are reserved for this later stage.

Rearrangement of Quaternary Ammonium Ions

The change (1) → (2) → (3) (*vide supra*) is effected, often in 90 per cent yield, by warming with aqueous or ethanolic alkali. The first, reversible, stage can be recognised independently of the actual rearrangement. The reaction is first-order in ammonium ion and is not accelerated by further molecules of alkali. The few observations at different temperatures indicate the high Arrhenius activation energy of about 130 kJ/mole. Since the process (2) → (3) involves the disappearance of formal charges, it naturally proceeds much faster in solvents of low dielectric constant [1].

Table 2. Migration Rates of Substituted Benzyl Groups

	Substituent in benzyl group					
	Cl	Br	I	NO_2	OMe	Me
o-	36	48	81	1040	1·9	15
m-	2·4	2·1	1·9	3·8	0·93	0·97
p-	2·6	2·8	3·2	73	0·76	1·06

Unsubstituted benzyl = 1·0.

Electron-attracting substituents in the aromatic nucleus of the benzyl group accelerate migration considerably (Table 2), in general accord with the suggestion of an internal S_N2 process. The striking effect of *ortho*-substitution has not been adequately discussed, and it is not known how far the effect is general for electrophilic rearrangements.

On the other hand, the reaction is retarded by electron-attracting substituents in the phenacyl ($PhCOCH_2$—) group. Under the usual reaction conditions, the rapid formation of the ylide (2) is assured; electron-attracting substituents tend to stabilise the anionic charge and so diminish the driving force for the rate-determining stage (2) → (3).

That the reaction is strictly *intra*molecular is shown by mixture experiments with labelled materials. In the case formulated, no radioactivity appeared in (4), and there is only a small difference (0·85:1) in the speed of rearrangement of the two ammonium ions [2].

$$\left.\begin{array}{c} p\text{-}BrC_6H_4COCH_2\overset{+}{N}Me_2(^{14}CH_2Ph) \\ PhCOCH_2\overset{+}{N}Me_2(CH_2Ph) \end{array}\right\} \xrightarrow{HO^-} \left\{\begin{array}{c} p\text{-}BrC_6H_4COCH(^{14}CH_2Ph)NMe_2 \\ PhCOCH(CH_2Ph)NMe_2 \end{array}\right.$$

(4)

The process is applicable to quite a wide range of structural analogues. Normally the stationary groups attached to nitrogen are alkyls, because alkylarylbenzylphenacylammonium ions are hard to obtain. As well as benzyl and substituted benzyl, the migrating group can be allyl, phenylpropargyl ($PhC{\equiv}CCH_2$—) (but see p. 109), or phenacyl; under the usual experimental conditions, simple alkyl or aryl groups do not migrate.

$$\begin{matrix} RO_2CCH_2Cl \\ + \\ PhCH_2NMe_2 \end{matrix} \longrightarrow \begin{matrix} RO_2CCH_2\overset{+}{N}Me_2 \\ | \\ CH_2Ph \end{matrix} \longrightarrow \begin{matrix} RO_2CCHNMe_2 \\ | \\ CH_2Ph \end{matrix} \longrightarrow \begin{matrix} HO_2CCHNMe_2 \\ | \\ CH_2Ph \\ (5) \end{matrix}$$

Wide variation is possible in the molecular environment of the destination carbon atom. The activating carbonyl group may be that of an ester function, as in the synthesis of *N*,*N*-dimethylphenylalanine (5). Non-hydroxylic conditions are necessary at the rearrangement stage to prevent ester hydrolysis.

Me$_2$N$^+$ H | PhCH$_2$ (6) $\xrightarrow{MeO^-}$ Me$_2$N CH$_2$Ph (7)

The hydrocarbon fluorene is almost as acidic as acetophenone, so it is not surprising that the reaction (6) → (7) proceeds quite easily.

When the migration terminus is less favourably situated, more energetic conditions are required to extract the proton, though the zwitterion (ylide) once formed will no doubt have a greater tendency to rearrangement. Thus the dibenzyldimethylammonium ion was first rearranged by fusion with sodamide.

$$\begin{matrix} PhCH_2\overset{+}{N}Me_2 \\ | \\ CH_2Ph \end{matrix} \xrightarrow{H_2N^-} \begin{matrix} Ph\overset{-}{C}H\overset{+}{N}Me_2 \\ | \\ CH_2Ph \end{matrix} \longrightarrow \begin{matrix} PhCHNMe_2 \\ | \\ CH_2Ph \\ (8) \end{matrix}$$

The same rearrangement was effected later by treatment with phenyl-lithium in ether (Wittig, [3]) or with sodamide in liquid ammonia (Hauser, [4]). Note that a solvent of low dielectric constant is advantageous, and that phenyl-lithium ranks as a very strong base indeed.

Sommelet reported [5] an unforeseen complication (Scheme 1). If

Scheme 1

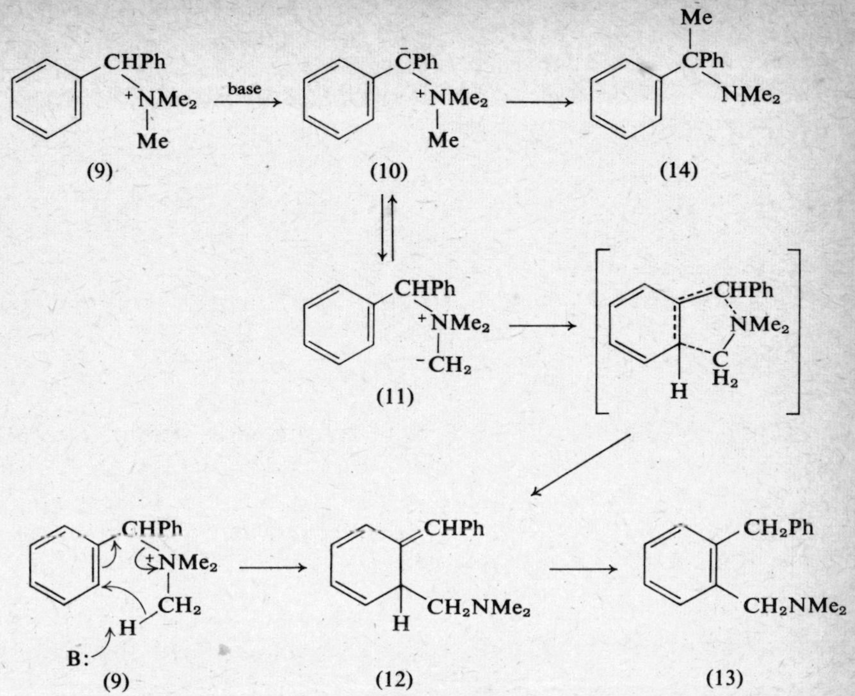

(9) (10) (14)

(11)

(9) (12) (13)

the hydroxide of (9) is kept in a warm place in a desiccator, it is converted into (13). Later, Wittig found that treatment with phenyl-lithium gives both (13) and the normal product (14) of 1,2-electrophilic rearrangement.

According to one view, the alternative highly reactive ylide (11) is formed in small quantity, and rearranges by a cyclic process in which the loss of benzene resonance is offset by the extinction of unfavourable formal charges. The product (12) naturally re-aromatises by prototropic change to (13). Another view suggests that the basic reagent takes a direct part in a one-stage production of (12).

With sodamide in liquid ammonia [4] or with phenyl-lithium [3], dibenzyldimethylammonium ion gives the Sommelet-type product, o-$MeC_6H_4CHPhNMe_2$, as well as the base (8) above. Here, and in the preceding case, high temperature favours the product of 1,2-migration (see p. 109).

The benzyltrimethylammonium ion, unlike phenacyl- or fluorenyl-trimethylammonium, and like ion (9), yields an ylide so basic as to undergo migration of *methyl* as well as the Sommelet change (Wittig).

The nature of the Sommelet change has been confirmed by a labelling experiment:

$$\mathrm{Ph^{14}CH_2\overset{+}{N}Me_3 \longrightarrow \textit{o}\text{-}^{14}CH_3C_6H_4CH_2NMe_2}$$

(15) → (16) [$NaNH_2$]

The type of quinonoid intermediate postulated as leading to (13) has been isolated in the case (15) → (16), in which its further transformation is blocked [6].

These rearrangements have found intriguing applications in synthesis. Hauser effected two Sommelet sequences (A [4], B).

(A) $C_6H_5CH_2\overset{+}{N}Me_3 \longrightarrow$ o-$MeC_6H_4CH_2NMe_2$ $\xrightarrow{1.\ MeI,\ 2.\ NaNH_2}$ 2,3-$Me_2C_6H_3CH_2NMe_2$ - - - → $C_6Me_5CH_2NMe_2$

(B) $C_6H_5CH\overset{+}{N}Me_2(CH_2)_n$ (cyclic) ⟶ benzo-fused ring with CH_2–$(CH_2)_n$–NMe–CH_2

Wittig [3] synthesised dibenzocyclo-octatetraene by a combination of 1,2-migrations and Hofmann degradations (Scheme 2). The by-product (17) was the first recorded unreduced isoindole.

Scheme 2

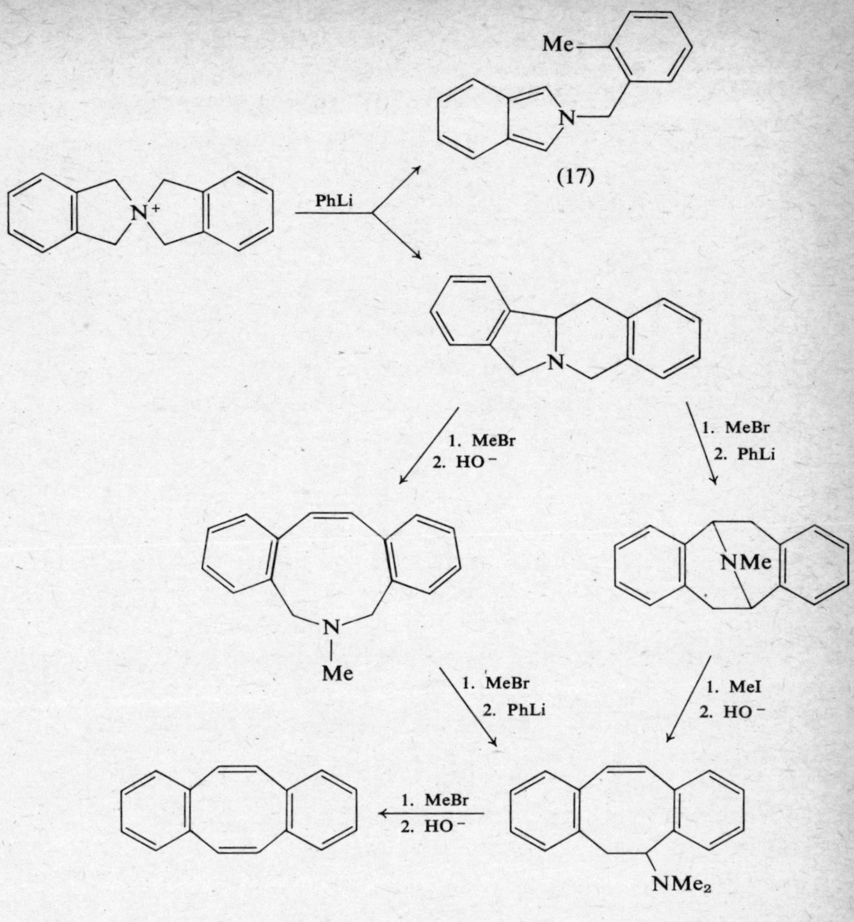

Me

N

(17)

N+

PhLi

N

1. MeBr
2. HO⁻

1. MeBr
2. PhLi

NMe

N

Me

1. MeBr
2. PhLi

1. MeI
2. HO⁻

1. MeBr
2. HO⁻

NMe_2

Several 1-benzyltetrahydroisoquinolines related to the simpler opium alkaloids have been prepared (Grethe), and the benzoquinolizine skeleton, basic to the alkaloid cryptopleurine, was produced (Stevens) by rearrangement of a cyclic allylammonium system.

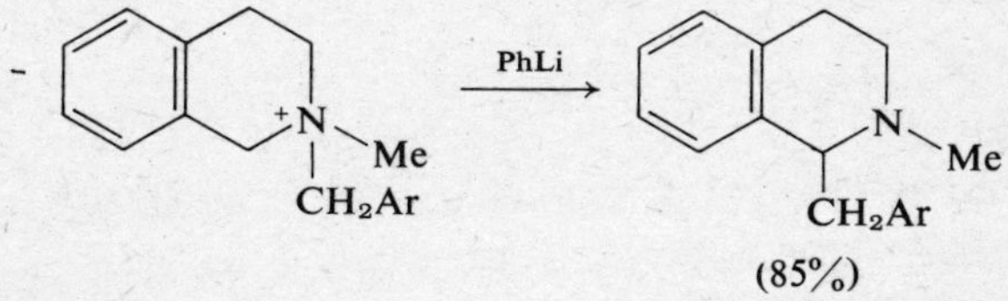

+N Me CH_2Ar —PhLi→ N Me CH_2Ar

(85%)

$$C_6H_4(CH_2Br)_2 + HN\text{(tetrahydropyridine)} \longrightarrow \text{spiro-}N^+ \xrightarrow[-H^+]{PhLi} \text{ylide} \longrightarrow \text{(benzoquinolizine)} \xrightarrow{H_2/Pd} \text{(saturated benzoquinolizidine)}$$

Rearrangement of Isocyanides

The thermal rearrangement of isocyanides, recorded a century ago by Weith, is formally a transformation of an ammonium ylide:

$$\overset{-}{C}\equiv\overset{+}{N}—R \xrightarrow{200^\circ} R—C\equiv N$$

More recently, with lessening fear of isocyanides, the reaction has been reinvestigated. It is a nearly quantitative, first-order process (R = alkyl or aryl) with energy of activation about 150 kJ/mole. The reaction is in the main intramolecular, since optically active sec. butyl isocyanide rearranges with substantial retention of configuration. On the other hand, the tertiary group, —CMeEtPh, migrates with predominant racemisation. The speed of isomerisation is little affected by *p*-substitution in an aryl group and somewhat diminished by chain-branching in alkyl [7]. It is not yet clear that this rearrangement fits into the general scheme developed later in this chapter.

Rearrangement of Ethers

Change from cationic, quadricovalent nitrogen to neutral, bicovalent oxygen as atom of departure does not prevent rearrangement. However, as would be expected, the initial extraction of a proton is much more difficult, and RO^- is a less favourable leaving-group than R_3N. Compare the sequence:

$$RCH_2O\text{—}R' \xrightarrow{-H^+} R\overset{-}{C}HO\text{—}R' \longrightarrow RCH(R')O^- \xrightarrow{H^+} RCH(R')OH$$

with the sequence:

$$RCH_2\overset{+}{N}Me_2\text{—}R' \xrightarrow{-H^+} R\overset{-}{C}H\overset{+}{N}Me_2\text{—}R' \longrightarrow RCH(R')NMe_2$$

While isolated observations are recorded earlier, we are indebted to Wittig for a systematic investigation [8] of this rearrangement. On treatment with phenyl-lithium in ether, fluorenyl ethers undergo rearrangement as shown.

$$\text{fluorenyl(H)(OR)} \xrightarrow{-H^+} \text{fluorenyl}^-\text{(OR)} \longrightarrow \text{fluorenyl(R)(O}^-) \xrightarrow{H^+} \text{fluorenyl(R)(OH)}$$

Reaction proceeds at 40° for R = alkyl or benzyl, at 100° for R = methyl, ethyl, or *p*-nitrophenyl, and fails for R = phenyl. This is in consonance with the diminishing ease of nucleophilic substitution of Br, for example, on the same groups R. Butanolic sodium 1-butoxide, a reagent of less extreme alkalinity, at 120° effected rearrangement of the benzyl and allyl but not alkyl or aryl fluorenyl ethers.

As in the ammonium ion series, the proton can be mobilised by an adjacent carbonyl group. Curtin studied [9] desyl ethers:

$$\text{PhCOCH(OR)Ph} \xrightarrow[\text{EtOH}]{\text{KOH}} \text{PhCOC(R)(OH)Ph}$$

The rate of this rearrangement varied with the nature of R in the descending order:

$$R = Ph_2CH > PhCH_2 > p\text{-}O_2NC_6H_4 > Ph$$

Rearrangement has also been observed when the mobilising group is CO_2Et or CN as shown by the following examples:

$$\text{PhCH(OCHPh}_2\text{)CO}_2\text{Et} \xrightarrow[\text{EtOH}]{\text{EtONa}} \text{PhC(CHPh}_2\text{)(OH)CO}_2\text{Et}$$

$$\underset{\displaystyle |\atop \displaystyle OCH_2Ph}{PhCHCN} \xrightarrow[BuOH]{BuONa} \left[\begin{array}{c} CH_2Ph \\ | \\ PhCCN \\ | \\ OH \end{array} \right] \longrightarrow PhCOCH_2Ph + HCN$$

(Note that cyanohydrins of aryl ketones are seldom stable.)

$$\begin{array}{ccc} \underset{(18)}{PhCOCH_2OCH_2Ph} & \xrightarrow{BuONa} & \begin{array}{c} OH \\ | \\ PhCCH_2Ph \\ | \\ CO_2H \\ (19) \end{array} \\ \downarrow & & \uparrow \\ \begin{array}{c} PhCOCHOH \\ | \\ CH_2Ph \end{array} & \xrightarrow{-2H} & \begin{array}{c} PhCOCO \\ | \\ CH_2Ph \end{array} \end{array}$$

Benzyl phenacyl ether (18) showed an unforeseen complication by yielding the acid (19). Stevens suggested the mechanism shown (1,2-migration; dehydrogenation; benzilic rearrangement). Equally plausible would be an explanation along the same lines as that offered for the change leading to (34) and (35) (p. 105).

Wittig's detailed study of benzhydryl ethers gave interesting results. Zinc and magnesium compounds of the type [$ROCPh_2$ (metal)] which are colourless and presumably covalent, do not rearrange. In contrast the corresponding compounds of the alkali metals, coloured and at least in part ionic, do rearrange. In dilute solutions, the rearrangement proceeds, contrary to natural expectation, with somewhat diminishing speed as the metal is changed from Li to Na to K; increased concentration increases the disparity in favour of Li. Wittig suggests that the smaller Li cation can more easily form a co-ordination complex, $Ph_2\bar{C}—\overset{\displaystyle R}{\overset{|}{O}} \rightarrow Li$, having essentially oxonium oxygen and hence a much greater tendency to rearrange.

With phenyl-lithium (Wittig) or with potassium amide (Hauser), benzyl methyl ether and dibenzyl ether gave moderate yields of the normal rearranged product, PhCHROH, while benzyl ethyl ether gave mainly benzyl alcohol and presumably ethylene (the usual fission of an ethyl ether by an active organometallic compound). Schöllkopf, using t-butyl-lithium, observed [10] rearrangement with R = Me,

Et, i-Pr, t-Bu, the rate increasing in that order. The ethyl ether gave a lower yield of PhCHEtOH, with much benzyl alcohol. He observed no 'scrambling' in a mixture experiment.

As well as undergoing a normal Wittig rearrangement, benzyl butynyl ether, $PhCH_2OCH_2C{\equiv}CMe$, suffers a Sommelet-type change (Schöllkopf), giving *o*-$MeC_6H_4CH(OH)C{\equiv}CMe$.

West observed [11] a reversible Wittig migration of a trialkylsilyl group:

$$>CHOSiR_3 \rightleftharpoons >C(OH)SiR_3$$

Wittig used this ether rearrangement to effect a serviceable synthesis of phenanthrene as shown.

$$\text{2,2'-}(CO_2Et)_2\text{-biphenyl} \xrightarrow{LiBH_4} \text{2,2'-}(CH_2OH)_2\text{-biphenyl} \xrightarrow{-H_2O} \text{dibenzoxepine} \rightarrow \text{9-hydroxy-9,10-dihydrophenanthrene} \xrightarrow{-H_2O} \text{phenanthrene}$$

Rearrangement of Tertiary Amines

This rearrangement, foreseeably harder to effect than that of ethers, has been little studied. The preliminary extraction of a proton is more difficult, and the change from uncharged to anionic nitrogen is unfavourable in comparison with the corresponding change for oxygen. The latter disadvantage is mitigated if the nitrogen atom carries an aryl group. Dahn encountered two unexpected results in the reduction of acylaminofluorenes (20).

$$\text{(20)}\ \text{fluorenyl: H, MeNCOC}_{10}\text{H}_7 \xrightarrow[\text{THF}]{\text{LiAlH}_4} \text{fluorenyl: MeNH, CH}_2\text{C}_{10}\text{H}_7$$

($C_{10}H_7$ = 1- and 2-naphthyl)

Stevens rearranged phenacylbenzylaniline by fusion with potassium hydroxide under mild conditions [12]:

$$\text{PhCOCH}_2\text{N(CH}_2\text{Ph)Ph} \xrightarrow[140°]{\text{KOH}} \text{PhCOCH(CH}_2\text{Ph)NHPh}$$

The reaction was shown to be intramolecular by a mixture experiment with labelled material leading to inactive (21).

$$\left.\begin{array}{c}\text{PhCOCH}_2\text{N(}^{14}\text{CH}_2\text{Ph)Ph} \\ + \\ t\text{-BuCOCH}_2\text{N(CH}_2\text{Ph)Ph}\end{array}\right\} \longrightarrow \left\{\begin{array}{c}\text{PhCOCH(}^{14}\text{CH}_2\text{Ph)NHPh} \\ + \\ t\text{-BuCOCH(CH}_2\text{Ph)NHPh}\end{array}\right.$$

(21)

Under very extreme conditions, dibenzylaniline and dibenzylmethylamine give respectively 8 per cent and 2 per cent of stilbene; the expected products of rearrangement break down to stilbene and primary base when subjected to the same drastic treatment:

$$(\text{PhCH}_2)_2\text{NR} \xrightarrow[260°]{\text{NaNH}_2} \begin{cases}\text{PhCH}_2\text{CHPhNHR} \longrightarrow \text{PhCH=CHPh} + \text{H}_2\text{NR} \\ \text{PhCH}_3 + \text{PhCH=NR}\end{cases}$$

The main process is an eliminative fission leading to a Schiff base.

An application of this rearrangement leads by ring contraction to a moderate yield of phenanthrene (cf. p. 91):

NPh $\xrightarrow[250^\circ]{NaNH_2}$ NHPh $\longrightarrow$ + H_2NPh

Rearrangement versus Eliminative Fission

Electrophilic rearrangement (A) and fission by elimination (B) appear to be alternative processes:

$$>CH-X(R) \xrightarrow{-H^+} >\bar{C}-X(R) \begin{cases} \xrightarrow{(A)} -C(R)-X^- \xrightarrow{H^+} -C(R)-XH \\ \xrightarrow{(B)} >C{=}X + R^- \xrightarrow{H^+} >C{=}X + RH \end{cases}$$

(The extraction of a proton is not regarded as necessarily independent of the following step.)

'Every schoolboy knows' the extreme case, where X = CH_2 and R = halogen and the normal reaction is elimination of HR by strong base. One could expect rearrangement to be favoured by tolerance of X for an unshared electron-pair, i.e. $\overset{+}{N}R_2$ > O > NPh > NMe, and the elimination by similar tolerance in R. These circumstances, however, have to be very favourable before rearrangement will occur (basicity of $X^- \ll$ basicity of R^-). Thus, systems of the type $(PhCH_2)_2X$ react as follows:

X = $\overset{+}{N}Me_2$	rearrangement
O	rearrangement + elimination
NPh	elimination + a little rearrangement
NMe	elimination + less rearrangement

The ether $Ph_2CHOCHPh_2$, not surprisingly, undergoes solely elimination, giving Ph_2CH_2 and Ph_2CO; the anion $R^- \equiv \bar{C}HPh_2$ is comparatively stable.

It is clear (Hauser) that rearrangement does not normally proceed by a direct process of fission followed by re-attachment:

$$\underset{\displaystyle R}{\overset{\displaystyle \bar{C}{-}X}{|}} \longrightarrow \underset{+R^-}{C{=}X} \longrightarrow \underset{\displaystyle R}{\overset{\displaystyle C{-}X^-}{|}} \xrightarrow{H^+} \underset{\displaystyle R}{\overset{\displaystyle C{-}XH}{|}}$$

This conclusion is drawn from the following observations:

(a) various mixture experiments show that R does not become kinetically free (pp. 83, 91, 92, 99);

(b) rearrangement of ethers can be effected in liquid ammonia, and of ammonium ions in water, either of which media would at once convert R^- into RH;

(c) rearrangement of $>$CHOR by R′Li gives, at most, only minor quantities of $>$CR′OH (but see pp. 95, 96).

Conversely, however, it is possible that in some exceptional cases, overall fission may proceed by way of rearrangement; that is:

$$\underset{\displaystyle R}{\overset{\displaystyle CH{-}X}{|}} \xrightarrow{-H^+} \underset{\displaystyle R}{\overset{\displaystyle \bar{C}{-}X}{|}} \longrightarrow \underset{\displaystyle R}{\overset{\displaystyle C{-}\bar{X}}{|}} \longrightarrow \underset{+R^-}{C{=}X} \xrightarrow{H^+} \underset{+RH}{C{=}X}$$

For example, the normal product of rearrangement of dibenzhydryl ether, Ph_2CHCPh_2OH, *may* be an intermediate step in the observed fission of the ether to diphenylmethane and benzophenone; these fission products are formed *more* rapidly from tetraphenylethanol than from the ether (Curtin).

Migration from Uncharged Carbon to Anionic Carbon

This rearrangement, on all considerations, should be the most difficult to effect. The first detailed studies are the simultaneous and independent work of Grovenstein [13] and Zimmermann [14]. To achieve the necessary driving force, the destination atom was not attached to any aryl or vinylic system capable of dispersing the anionic charge; the carbanion was produced by reaction of an alkyl halide function with an alkali metal:

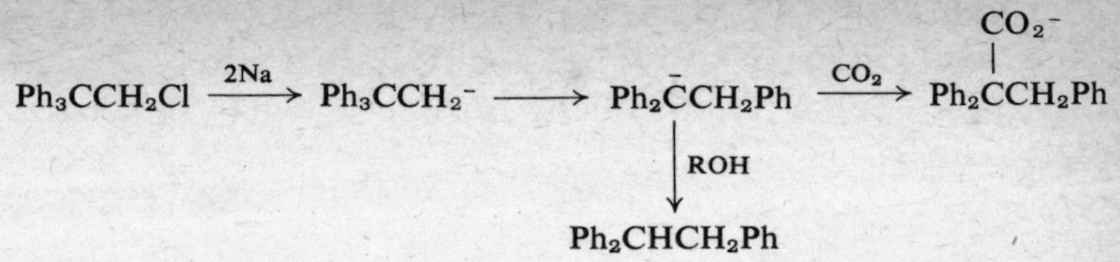

$$Ph_3CCH_2Cl \xrightarrow{2Na} Ph_3CCH_2^- \longrightarrow Ph_2\bar{C}CH_2Ph \xrightarrow{CO_2} Ph_2C(CO_2^-)CH_2Ph$$

$$Ph_2\bar{C}CH_2Ph \xrightarrow{ROH} Ph_2CHCH_2Ph$$

A rather striking correlation between the hetero-polar character of an organometallic compound and the tendency of the system to undergo rearrangement is shown by experiments on organometallic reactants derived from 2,2-diphenylpropyl chloride (22) (Scheme 3).

Scheme 3

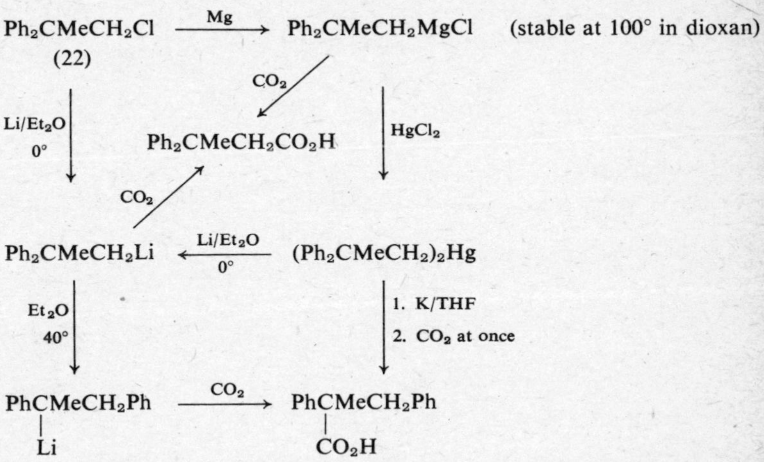

Ph_2CMeCH_2Cl (22) $\xrightarrow{Mg}$ Ph_2CMeCH_2MgCl (stable at 100° in dioxan)

$Ph_2CMeCH_2MgCl \xrightarrow{CO_2} Ph_2CMeCH_2CO_2H$

$Ph_2CMeCH_2MgCl \xrightarrow{HgCl_2} (Ph_2CMeCH_2)_2Hg$

$Ph_2CMeCH_2Cl \xrightarrow{Li/Et_2O,\ 0°} Ph_2CMeCH_2Li$

$(Ph_2CMeCH_2)_2Hg \xrightarrow{Li/Et_2O,\ 0°} Ph_2CMeCH_2Li$

$Ph_2CMeCH_2Li \xrightarrow{CO_2} Ph_2CMeCH_2CO_2H$

$Ph_2CMeCH_2Li \xrightarrow{Et_2O,\ 40°} PhCMe(Li)CH_2Ph$

$(Ph_2CMeCH_2)_2Hg \xrightarrow{1.\ K/THF;\ 2.\ CO_2\ at\ once} PhCMe(CO_2H)CH_2Ph$

$PhCMe(Li)CH_2Ph \xrightarrow{CO_2} PhCMe(CO_2H)CH_2Ph$

Note that alkylmercury compounds, in general, are strictly covalent, Grignard compounds substantially so, and even alkyl-lithiums are mainly covalent.

There is an embarrassing *arrière-pensée* regarding these rearrangements. By the action of an alkali metal, a compound RHal can yield not only a carbanion R^-, but also a free radical $R\cdot$. The changes we have been considering could be radical rearrangements. Against this possibility it may be adduced that the results of carbonation in the experiments with (22) show the presence of carbanions, and that, in compounds of the type $ArAr'_2CCH_2Li$, phenyl migrates in preference to *p*-tolyl.

A further complication has been observed in the reaction of $PhCH_2CPh_2CH_2Cl$ with lithium, which gives $Ph_2CHCH_2CH_2Ph$ by migration of benzyl. If rearrangement is carried out in presence of labelled benzyl-lithium ($Ph^{14}CH_2Li$), the product is extensively labelled, indicating a full dissociation-recombination sequence:

$$PhCH_2CPh_2CH_2Li \longrightarrow \begin{matrix} PhCH_2Li \\ + \\ Ph_2C{=}CH_2 \end{matrix} \longrightarrow \begin{matrix} Ph_2CCH_2CH_2Ph \\ | \\ Li \end{matrix}$$

Benzyl-lithium does, in fact, combine in this way with 1,1-diphenylethylene. On the other hand, a parallel experiment with Ph_3CCH_2Li and labelled PhLi showed no uptake of labelled phenyl.

Hauser reports a case in which the required carbanion is produced by the more accustomed device of proton-extraction:

$$Ph_3CCH_2Ph \xrightarrow[100^\circ]{PhNa} Ph_2CHCHPh_2$$

Molecular-orbital calculations are consistent with the fact that, in this series, phenyl migrates faster than methyl. Qualitatively, in a case like this where the ultimate anionic carbon atom is an unfavourable site for the charge, rearrangement will be promoted if the migrating-group can accommodate a substantial share of negative charge in the transition state (23).

H_2C^- / R_2C ⟷ H_2C / R_2C ⟷ H_2C / R_2C^-

(23)

The observed migration of phenyl in preference to *p*-tolyl is well accounted for. This species (23) is, in fact, a parallel to the Meisenheimer complexes concerned in nucleophilic substitution in aromatic systems, and may represent a discrete intermediate rather than merely a transition state.

Rearrangement of 2,2-Diarylvinyl Halides (Fritsch)

This transformation forms part of a synthetically useful sequence leading to diarylacetylenes. The mechanism is elucidated by a case in which the initial ketone is isotopically labelled.

Configurations were assigned to the diarylvinyl bromides by measurement of dipole moments. When the final acetylenes were oxidised, greatly predominating radioactivity was associated in the one case

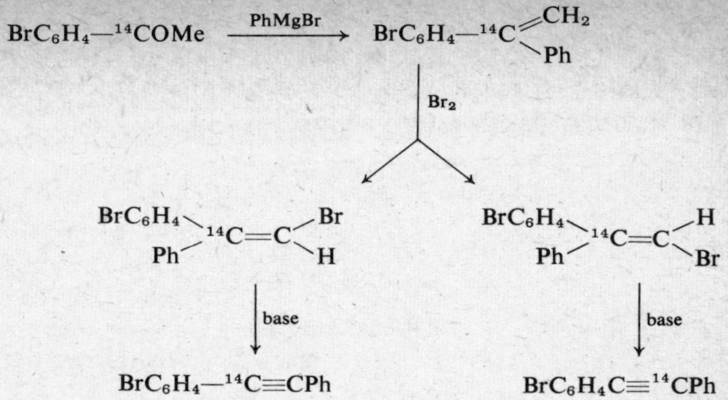

$$\mathrm{BrC_6H_4{-}^{14}COMe} \xrightarrow{\mathrm{PhMgBr}} \mathrm{BrC_6H_4{-}^{14}C({=}CH_2)Ph}$$

$\xrightarrow{\mathrm{Br_2}}$

$$\mathrm{(BrC_6H_4)(Ph)^{14}C{=}C(Br)(H)} \xrightarrow{\text{base}} \mathrm{BrC_6H_4{-}^{14}C{\equiv}CPh}$$

$$\mathrm{(BrC_6H_4)(Ph)^{14}C{=}C(H)(Br)} \xrightarrow{\text{base}} \mathrm{BrC_6H_4C{\equiv}^{14}CPh}$$

with *p*-bromobenzoic acid, in the other with benzoic acid. It follows that the carbene, $Ar_2C{=}C{:}$, in which the stereochemical distinction disappears, is not an intermediate. Removal of a proton from the vinyl bromide would give the anion,

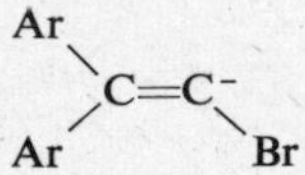

$$\mathrm{Ar_2C{=}C^{-}{-}Br}$$

and separation of Br^- with migration of *trans*-Ar would find a significant parallel in the Beckmann change.

Synthetically, diarylvinyl chlorides or bromides may be treated with butyl-lithium or with sodium or potassium amide in liquid ammonia.

Rearrangement of Hydrazine and Hydroxylamine Derivatives

Wawzonek effected [15] a sequence involving the ylide (25). It behaves as a weak base, neutral to phenolphthalein, easily interconvertible with (24). There is thus no great driving force for the rearrangement but no need for a further alkaline reagent.

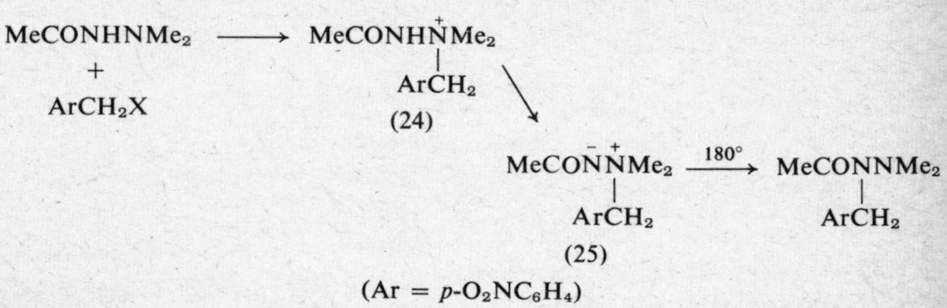

$$\mathrm{MeCONHNMe_2 + ArCH_2X} \longrightarrow \mathrm{MeCONH\overset{+}{N}Me_2(CH_2Ar)}\ (24) \longrightarrow \mathrm{MeCO\overset{-}{N}\overset{+}{N}Me_2(CH_2Ar)}\ (25) \xrightarrow{180^\circ} \mathrm{MeCON(CH_2Ar)NMe_2}$$

$(Ar = p\text{-}O_2NC_6H_4)$

When the hydrazine system is non-quaternary and not acylated, the picture is very different [16]. It is now necessary to use enough of a powerful reagent to effect a double deprotonation, and then rearrangement takes place at a low temperature:

$$Ar_2NNH_2 \xrightarrow{RLi} \underset{\text{(stable)}}{Ar_2N\bar{N}H} \longrightarrow Ar_2NN^{2-} \longrightarrow Ar\bar{N}\bar{N}Ar \longrightarrow ArNHNHAr$$

The process is strictly intramolecular. It succeeds with $RArNNH_2$ (which group migrates?) but not with R_2NNH_2 (R = alkyl).

West has observed [16] very easy migration of trimethylsilyl groups in derivatives of hydrazine and hydroxylamine:

$$\begin{matrix} Me_3SiNH \\ | \\ Me_3SiNMe \end{matrix} \xrightarrow{BuLi} \begin{matrix} Me_3SiN^- \\ | \\ Me_3SiNMe \end{matrix} \longrightarrow \begin{matrix} (Me_3Si)_2N \\ | \\ {}^-NMe \end{matrix} \longrightarrow \begin{matrix} (Me_3Si)_2N \\ | \\ NHMe \end{matrix}$$

$$Me_3SiNHOSiMe_3 \xrightarrow{-H^+} Me_3Si\bar{N}OSiMe_3 \rightleftharpoons (Me_3Si)_2NO^-$$

The equilibrated anions were recognised by methylation, giving respectively $Me_3SiN(Me)OSiMe_3$ and $(Me_3Si)_2NOMe$.

A similar O → N migration of a benzhydryl group has been reported [17]:

$$Ph_2CHONHTs \xrightarrow{base} Ph_2CHO\bar{N}Ts \xrightarrow{165^\circ} \begin{matrix} Ph_2CHNTs \\ | \\ O^- \end{matrix} \longrightarrow \begin{matrix} Ph_2C{=}NOH \\ + Ts^- \end{matrix}$$

Alternatively, this process could be regarded as a rearrangement of an intermediate nitrenoid species ($Ph_2CHO\ddot{N}$:).

Rearrangement of Amine Oxides

Meisenheimer observed [18] that *N*-allyl-*N*-methylaniline oxide and *N*-benzyl-*N*-methylaniline oxide (26) are rearranged by distillation in steam from alkaline solution:

$$\underset{(26)}{\begin{matrix} Ph \\ | \\ Me\overset{+}{N}{-}\bar{O} \\ | \\ CH_2Ph \end{matrix}} \xrightarrow{100^\circ} \begin{matrix} Ph \\ | \\ MeN{-}OCH_2Ph \end{matrix}$$

Saturated alkyl groups do not migrate under these conditions. Amine oxides are weak ($pK_a \approx 4$) water-soluble bases, and the reaction does not require alkali in excess of that needed to liberate the amine oxide from a salt. The *N*-aryl group in the compounds studied by Meisenheimer is of course favourable to the reaction, which is nevertheless rather slow on account of the weak basicity of the recipient anionic oxygen.

Benzyl- and allyl-dialkylamine oxides rearrange when heated in the solid state (Cope). Benzhydryldialkylamine oxides rearrange very smoothly under mild conditions (i.e., $\bar{O}$—$\overset{+}{N}Me_2CHPh_2 \rightarrow Ph_2CHONMe_2$); mixture experiments show the change to be at least predominantly intramolecular (Stevens). On the other hand, some 'scrambling' has been reported in the rearrangement of a mixture of the *N*-oxides of *N*-benzylpiperidine and *N*-*p*-methylbenzylpyrrolidine [19].

Table 3. Migratory Aptitudes of Substituted Benzyl Groups in (26)

	Substituent in benzyl group				
	F	Cl	Br	NO_2	Me
m-	1·6	3·0	1·7	—	1·0
p-	0·74	3·3	—	32·7	0·58

Rate for unsubstituted benzyl = 1·0.

Substituents in the migrating benzyl group of (26) affect the rate of migration (Table 3) [20] in much the same way as in benzylphenacylammonium ions (p. 83); the Arrhenius energy of activation is about 150 kJ/mole.

The amine oxide rearrangement has been applied synthetically [21] (Quinn) to the expansion of pyrrolidines to tetrahydro-oxazines and of piperidines to hydro-oxazepines:

$$\text{ArCH—}\overset{+}{N}\text{Me}(\text{—}C_n\text{— ring})\text{, }N\text{—}O^- \longrightarrow \text{ArCH—O—NMe}(\text{—}C_n\text{— ring}) \qquad (n = 3 \text{ or } 4)$$

Nitrones. The *N*-oxides of Schiff bases are resonance-stabilised and rearrange less readily than other amine oxides:

$$\left[Ph_2\overset{+}{C}\text{—N—O}(\text{N—CHPh}_2) \longleftrightarrow Ph_2C\text{=}\overset{+}{N}\text{—}\bar{O}(\text{N—CHPh}_2) \right] \longrightarrow Ph_2C\text{=N—O—CHPh}_2$$

The reaction is first-order at 130–160° in diethylene glycol diethyl ether as solvent with an activation energy of about 160 kJ/mole; the rate is relatively insensitive to substitution in the phenyl groups [22].

New Routes to Ylides

(a) A carbene can combine with an atom carrying a lone pair of electrons [23]:

$$\text{9-diazofluorene } (=N_2) \xrightarrow[-N_2]{100^\circ} \text{fluorenylidene } (\ddot{C}) \xrightarrow{PhCH_2NMe_2} \text{fluorenyl}^{-}\text{—}\overset{+}{N}Me_2CH_2Ph \longrightarrow \text{9-}(Me_2N)\text{-9-}(CH_2Ph)\text{fluorene}$$

Irradiation of methyl diazomalonate in presence of dimethyl sulphide gives, no doubt by way of a carbene, an isolable ylide which can rearrange (Ando):

$$(MeO_2C)_2C{=}N_2 + SMe_2 \xrightarrow{h\nu} (MeO_2C)_2C^{-}\text{—}{}^{+}SMe_2 \xrightarrow{230^\circ} (MeO_2C)_2CMe\text{—}SMe$$

Formally similar reactions with dimethyl ether and with allyl bromide give analogous products, presumably via non-isolable oxonium and bromonium $[(MeO_2C)_2\bar{C}\overset{+}{Br}CH_2CH{=}CH_2]$ ylides. Dichlorocarbene has been applied similarly (Parham):

$$Cl_2C\colon + Et_2NCH_2CH{=}CMe_2 \longrightarrow Cl_2\bar{C}\text{—}\overset{+}{N}Et_2CH_2CH{=}CMe_2$$

$$\downarrow$$

$$Et_2NCOCH_2CH{=}CMe_2 \xleftarrow{H_2O} Et_2N\text{—}CCl_2CH_2CH{=}CMe_2$$

(b) Production of benzyne in presence of a compound with a lone-pair atom can also afford an ylide [24]:

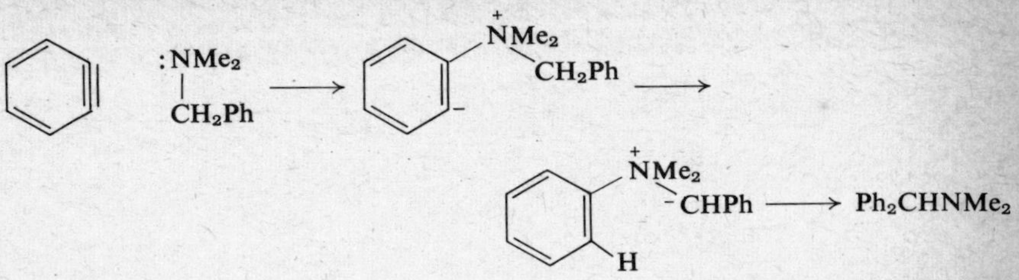

(benzyne) + $:NMe_2$–CH_2Ph ⟶ (carbanion, $\overset{+}{N}Me_2$, CH_2Ph) ⟶

($\overset{+}{N}Me_2$, $^{-}CHPh$, H) ⟶ Ph_2CHNMe_2

In the above sequence, note the migration of a phenyl group in the last step. This method has been used extensively with tertiary amines and with thio-ethers (*vide infra*).

Stereochemistry and the Migrating Anion Theory

In several cases it was found that rearrangements, in which the α-carbon atom of the migrating-group is a centre of asymmetry, proceeded with substantially complete retention of configuration. This is strong evidence that the changes are strictly intramolecular, since liberation of a kinetically free, chiral, 1-phenylethyl group, for example, whether as anion, cation, or free radical (i.e., $Ph\bar{C}HMe$, $Ph\overset{+}{C}HMe$, or $Ph\dot{C}HMe$), would lead to racemisation. Two cases may be cited (see Scheme 4 [25] and Scheme 5 [26]).

Scheme 4

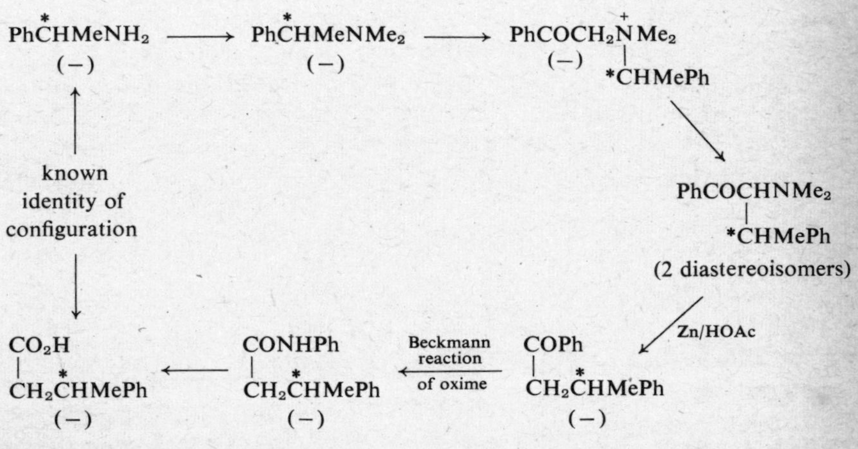

$Ph\overset{*}{C}HMeNH_2$ (−) ⟶ $Ph\overset{*}{C}HMeNMe_2$ (−) ⟶ $PhCOCH_2\overset{+}{N}Me_2$–*CHMePh (−)

⟶ $PhCOCHNMe_2$–*CHMePh (2 diastereoisomers)

$\xrightarrow{Zn/HOAc}$ $COPh$–$CH_2\overset{*}{C}HMePh$ (−) $\xrightarrow{\text{Beckmann reaction of oxime}}$ $CONHPh$–$CH_2\overset{*}{C}HMePh$ (−) ⟶ CO_2H–$CH_2\overset{*}{C}HMePh$ (−)

CO_2H–$CH_2\overset{*}{C}HMePh$ (−) ⟷ known identity of configuration ⟷ $Ph\overset{*}{C}HMeNH_2$ (−)

Scheme 5

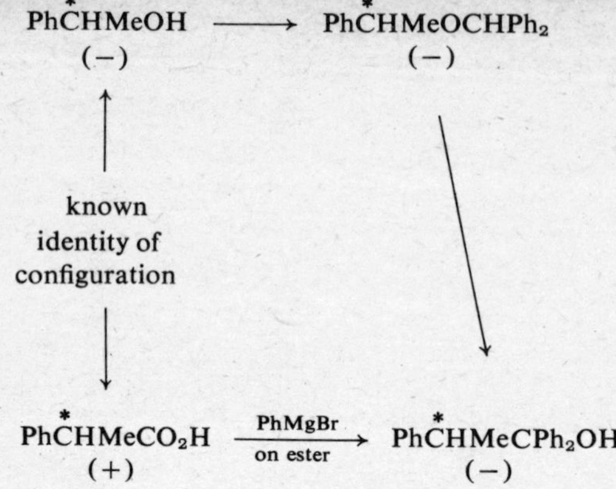

On the interpretation of the mechanism as an internal S_N2 reaction, retention of configuration at the migrating atom is in accord with a 'triangular' transition state, for example:

$$\underset{\text{PhCHMe}}{\overset{\delta-}{\text{PhCOCH}}\text{---}\overset{\delta+}{\text{NMe}_2}}$$

Such a process would involve frontal attack by C^- on the asymmetric carbon atom. This comfortable view was vigorously opposed in the early 1960s (Jenny, Lansbury, Schöllkopf). Such a concerted frontal attack is 'forbidden' on considerations of orbital symmetry (see Chapter 8). Moreover, cases accumulated in which rearrangement takes place with extensive racemisation. This was observed [27], for example, with benzyl sec. butyl ether, and with the chiral amine oxide, $\bar{O}\overset{+}{N}MePhCHDPh$.

$$\underset{(27)}{H_2C{=}CH{-}CH_2\overset{+}{N}Me_2(\text{*CHMePh})} \xrightarrow{NaNH_2} H_2\bar{C}{\cdots}CH{\cdots}CH\overset{+}{N}Me_2(\text{*CHMePh}) \longrightarrow \underset{(28)}{H_2C{=}CHCHNMe_2(\text{*CHMePh})} \;+\; \underset{(29)}{H_2C(\text{*CHMePh}){-}CH{=}CHNMe_2}$$

The rearrangement [28] of the ion (27), in which both a 1,2-shift (28) and a 1,4-shift (29) takes place, appeared very significant. Both products show retention of configuration; the internal S_N2 theory would, however, predict retention for (28) but inversion for (29).

If the idea of a one-stage concerted process must be abandoned, it seems that, in view of the degree of intramolecularity usually observed in these rearrangements, we must postulate a two-stage process, such that the migrating entity is detached and then re-attached, all within a solvent cage, without becoming kinetically free. There would appear to be three possibilities: (a) detachment of the migrating-group as cation; (b) detachment as anion; (c) detachment as free radical.

Proposal (a) is clearly unacceptable. First, it is difficult to see why the production of an anionic carbon atom by strong alkali should be a necessary preliminary to the ejection of a positively charged fragment; secondly, the favourable effect of electron-attracting substituents in the migrating-group is hardly consistent with its separation as cation; thirdly, groups, which as cation would undergo prompt nucleophilic rearrangement, here migrate unchanged.

The anion-migration theory (b) was asserted very convincingly. It accounts for the favourable influence of electron-attracting substituents in a migrating-group. The degree of retention of asymmetry in the migration of, for example, a chiral RR′CH group will be variable, according to the time that the anion spends in the solvent cage before reattachment. The residue remaining after the separation of an anion is usually a relatively stable entity, already known to be attacked by nucleophiles at the destination point of the rearrangement. Thus, for an ether we have:

$$\begin{array}{c}\mathrm{Ph\bar{C}H{-}O}\\ |\\ \mathrm{CH_2Ph}\end{array} \longrightarrow \left[\begin{array}{c}\mathrm{PhCH{=}O}\\ \\ \mathrm{\bar{C}H_2Ph}\end{array}\right] \longrightarrow \begin{array}{c}\mathrm{PhCH{-}O^-}\\ |\\ \mathrm{CH_2Ph}\end{array}$$

and for an amine:

$$\begin{array}{c}\mathrm{Ph\bar{C}H{-}NPh}\\ |\\ \mathrm{CH_2Ph}\end{array} \longrightarrow \left[\begin{array}{c}\mathrm{PhCH{=}NPh}\\ \\ \mathrm{\bar{C}H_2Ph}\end{array}\right] \longrightarrow \begin{array}{c}\mathrm{PhCH{-}\bar{N}Ph}\\ |\\ \mathrm{CH_2Ph}\end{array}$$

and for the ammonium ion:

$$\begin{array}{c}\mathrm{Ph\bar{C}H{-}\overset{+}{N}Me_2}\\ |\\ \mathrm{CH_2Ph}\end{array} \longrightarrow \left[\begin{array}{c}\mathrm{PhCH{=}\overset{+}{N}Me_2}\\ \\ \mathrm{\bar{C}H_2Ph}\end{array}\right] \longrightarrow \begin{array}{c}\mathrm{PhCH{-}NMe_2}\\ |\\ \mathrm{CH_2Ph}\end{array}$$

From general experience, the order of diminishing activity towards nucleophiles is: $>C{=}\overset{+}{N}R_2 > \ >C{=}O > \ >C{=}NR$. The picture for an amine oxide is plausible, if less supported by experience:

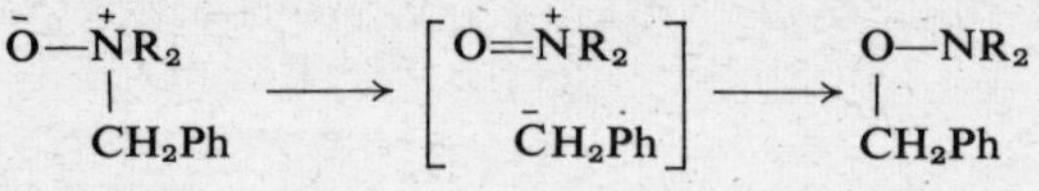

$$\bar{O}{-}\overset{+}{N}R_2(CH_2Ph) \longrightarrow \left[O{=}\overset{+}{N}R_2 \quad \bar{C}H_2Ph \right] \longrightarrow O{-}NR_2(CH_2Ph)$$

The first stage 'looks' less favourable than that for an ammonium ylide, and an amine oxide does in fact rearrange less readily than an ylide once formed.

Cases of 1,2- and 1,4-shift reported by Jenny can be accommodated. The rearrangement (27) → (28) and (29) (p. 102) would involve the intermediate

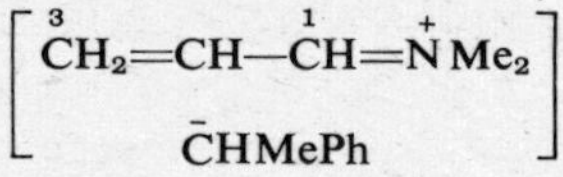

$$\left[\overset{3}{C}H_2{=}CH{-}\overset{1}{C}H{=}\overset{+}{N}Me_2 \quad \bar{C}HMePh \right]$$

with alternative attachment at C_1 and C_3.

Jenny also reports [28] a more elaborate example with an isotopically labelled carbon atom (30). Note the easy hydrolysis of the vinylamine (31) and the immonium ion (32) and that (33) → (34) parallels

Scheme 6

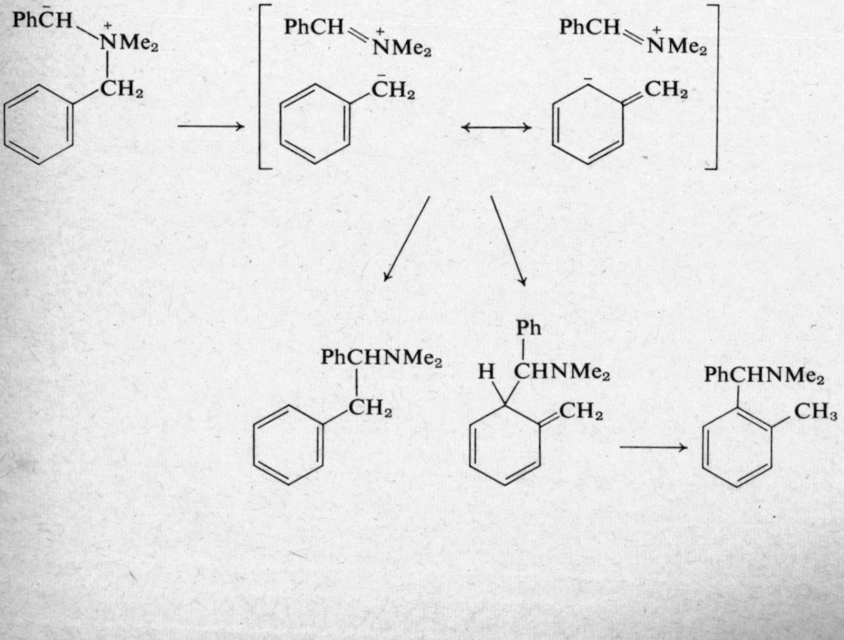

$$Ph\bar{C}H{-}\overset{+}{N}Me_2{-}CH_2C_6H_5 \longrightarrow \left[PhCH{=}\overset{+}{N}Me_2 \quad \bar{C}H_2C_6H_5 \longleftrightarrow PhCH{=}\overset{+}{N}Me_2 \quad CH_2{=}C_6H_5^{-} \right]$$

$$\swarrow \qquad \searrow$$

$$PhCHNMe_2{-}CH_2C_6H_5 \qquad \text{H, CHPhNMe}_2\text{-cyclohexadiene}{=}CH_2 \longrightarrow o\text{-}(PhCHNMe_2)C_6H_4CH_3$$

the glucose → fructose change. The end points (34) and (35) are identical chemically but differently labelled. The activity of the product corresponds to 43 per cent of (34) and 57 per cent of (35). Cram has pointed out that this theory also accommodates the Sommelet rearrangement very plausibly (see Scheme 6).

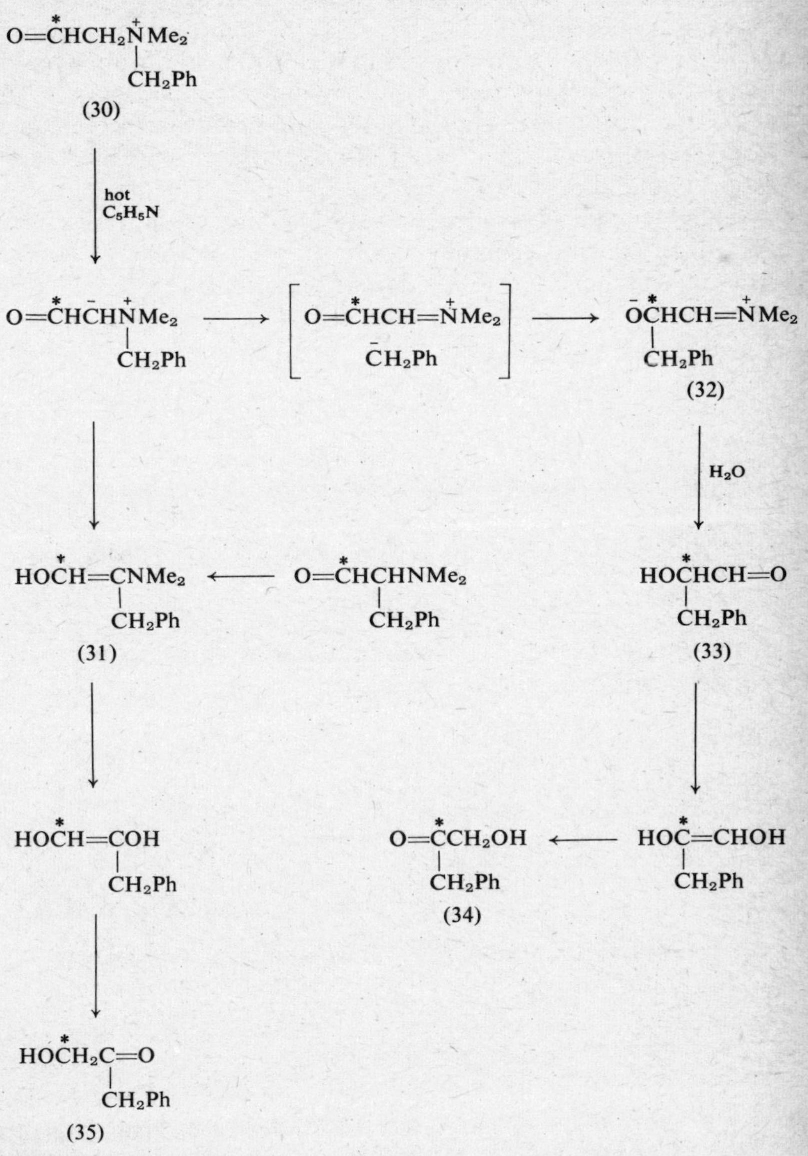

$O{=}\overset{*}{C}HCH_2\overset{+}{N}(CH_2Ph)Me_2$ (30)

↓ hot C_5H_5N

$O{=}\overset{*}{C}H\overset{-}{C}H\overset{+}{N}Me_2$ (CH_2Ph on N) ⟶ $[O{=}\overset{*}{C}HCH{=}\overset{+}{N}Me_2 \quad \overset{-}{C}H_2Ph]$ ⟶ $\overset{-}{O}\overset{*}{C}H(CH_2Ph)CH{=}\overset{+}{N}Me_2$ (32)

↓ (from first intermediate)

$HO\overset{*}{C}H{=}C(CH_2Ph)NMe_2$ (31) ⟵ $O{=}\overset{*}{C}HCH(CH_2Ph)NMe_2$

(32) ↓ H_2O

$HO\overset{*}{C}H(CH_2Ph)CH{=}O$ (33)

(31) ↓

$HO\overset{*}{C}H{=}C(CH_2Ph)OH$

↓

$HO\overset{*}{C}H_2C(CH_2Ph){=}O$ (35)

(33) ↓

$HO\overset{*}{C}(CH_2Ph){=}CHOH$ ⟶ $O{=}\overset{*}{C}(CH_2Ph)CH_2OH$ (34)

The Migrating Radical Theory

This third possibility, envisaging an intermediate radical-pair, was probably long in receiving adequate consideration because it seemed unnatural to look for such a mechanism for a reaction often initiated by drastic treatment with a powerful base—which almost irresistibly, though not logically, suggests a heterolytic process.

Some observations regarding the relative migratory aptitudes of groups were seen to be more consistent with a radical fission. Thus, in the phenacylammonium series, the 1-phenylethyl group migrates faster than benzyl by three powers of ten, and in the Wittig rearrangement of alkyl benzyl ethers, tertiary alkyl groups migrate most easily (Lansbury; Schöllkopf [10]).

It is enlightening to consider the nature of the second radical concerned. In the case of a Meisenheimer rearrangement, this would be of a well known stable type, an alkylaryl-nitrogen oxide:

$$\overset{-}{O}\text{—}\overset{+}{N}MePh(CH_2Ph) \longrightarrow \left[:\ddot{O}^{-}\text{—}\dot{\overset{+}{N}}MePh \longleftrightarrow :\ddot{\dot{O}}\text{—}\ddot{N}MePh \quad \dot{C}H_2Ph \right] \longrightarrow O\text{—}NMePh(CH_2Ph)$$

The corresponding entity in a Wittig rearrangement would be the radical-anion of the stable 'metal-ketyl':

$$Ph_2\overset{-}{C}\text{—}O(CH_2Ph) \longrightarrow \left[Ph_2\overset{-}{\ddot{C}}\text{—}\dot{\ddot{O}}: \longleftrightarrow Ph_2\dot{C}\text{—}\ddot{\ddot{O}}:^{-} \quad \dot{C}H_2Ph \right] \longrightarrow Ph_2C(CH_2Ph)\text{—}O^{-}$$

For the rearrangement of ammonium ions, the picture is less immediately satisfactory:

$$Ph\overset{-}{C}H\text{—}\overset{+}{N}Me_2(CH_2Ph) \longrightarrow \left[Ph\overset{-}{\ddot{C}}H\text{—}\dot{\overset{+}{N}}Me_2 \longleftrightarrow Ph\dot{C}H\text{—}\ddot{N}Me_2 \quad \dot{C}H_2Ph \right] \longrightarrow PhCH(CH_2Ph)\text{—}NMe_2$$

We have less information as to the stability of radicals such as

$$\text{—}\dot{C}\text{—}\ddot{N}\text{—} \longleftrightarrow \text{—}\overset{-}{\ddot{C}}\text{—}\dot{\overset{+}{N}}\text{—}$$

but it may be remembered that, in tertiary amines (and in ethers), various 'homolytic-looking' processes, including autoxidation, attack the bond

$$H \,\vdots\, C\text{—}N\text{—} \quad \text{or} \quad H \,\vdots\, C\text{—}O\text{—}.$$

There is no lack of indication of the escape of radicals from the solvent cage. Thus, the rearrangement of benzylphenacylammonium ions gives minor quantities of 1,2-diphenylethane [29].

$$PhCO\bar{C}H{-}\overset{+}{N}Me_2(CH_2Ph) \longrightarrow \left[PhCO\ddot{C}H{-}\dot{\overset{+}{N}}Me_2 \quad \dot{C}H_2Ph \right] \longrightarrow PhCH_2CH_2Ph$$

The corresponding sulphonium ion gives substantial quantities of $[PhCOCH(SMe)]_2$, as well as much diphenylethane. In the rearrangement of $MePhN^+(O^-)CH_2Ph$, the stable radical MePhNO can be recognised, by electron spin resonance spectroscopy [20], as a reaction product increasing with time—not strictly established as an intermediate in the formation of $MePhNOCH_2Ph$.

$$Me_3CCH_3 \longleftarrow Me_3CCH_2\overset{+}{N}Me_2(O^-) \longrightarrow Me_3CCH_2ONMe_2$$

(36)

Dimethylneopentylamine oxide (36) gives some neopentane as well as the product of migration of neopentyl. That neopentyl shifts, without transformation into tertiary amyl, is further strong evidence against a migrating cation theory (cf. Chapter 2). The nitrone $Ph_2C{=}N^+(O^-)CHPh_2$ gives [22] on rearrangement some tetraphenylethane $Ph_2CHCHPh_2$. Benecke and Wikel trapped free benzyl radicals in the rearrangement of, *inter alia*, $PhCH_2\overset{+}{N}Me_2\bar{N}COMe$.

Measurements of n.m.r., carried out in rapid succession during the course of a reaction, have been enlightening. When a product has just been formed by a radical reaction, the populations of the two magnetic levels of a contained proton can differ from the normal distribution, with the result that a characteristic band of the new-born product may register as an emission or an abnormal absorption. In contemporary style, the phenomenon is baptised C.I.D.N.P. (i.e., Chemically Induced Dynamic Nuclear Polarization) [30].

Such observations have indicated (1969) that radicals play a part in the rearrangement of the following, among other structures:

$PhCO\bar{C}H\overset{+}{N}Me_2CH_2Ph$ [29]; $PhCO\bar{C}H\overset{+}{S}MeCH_2Ar$ [29];
$PhCO\bar{C}H\overset{+}{S}MeCHDPh$ (Baldwin);
$MeCO\bar{N}\overset{+}{N}Me_2CH_2CH{=}CMe_2$ [31] (Morris); $PhCH_2NMe_2$-benzyne (Lepley);
$(PhCH_2)_2S$-benzyne (Iwamura).

Schöllkopf has maintained that the effect of electron-attracting substituents in increasing the migratory aptitude of a benzyl group (pp. 83, 99), though considerable, is much smaller than would be expected for actual separation of a benzyl anion and more in accord with the detachment of a free radical.

The migration of an aryl group, which is rather less reluctant than might be expected, is thought [13] to involve, exceptionally, the type of spirocyclic intermediate shown on p. 96. In the rearrangement of Ar_3CCH_2Cl in the presence of lithium, *p*-biphenylyl migrates 24·5 times as fast as *m*-biphenylyl, consonantly with participation of the ion

Ar_2C / H_2C $\longleftrightarrow$ Ar_2C / H_2C

Allylic Changes. Rearrangement of Sulphur Compounds

'Inversion' that accompanies displacement reactions in allylic systems is familiar, and the behaviour of allyl and related groups in this series of rearrangements presents some complexities. These are most evident, and have special points of interest, in sulphonium ions, consideration of which has in general been deferred to this section.

Considering first the ammonium ion series, we can envisage four possibilities. First, let an allyl group include the 'destination atom': we have processes A and B. Second, allyl as migrating-group presents possibilities C and D.

$RCH{=}CH{-}CH_2\overset{+}{N}Me_2R'$

$\downarrow$

$R\overline{CH{-}CH{-}CH}\overset{+}{N}Me_2R'$

$\nearrow$ $RCH{=}CHCHR'NMe_2$ (A)

$\searrow$ $RR'CHCH{=}CHNMe_2$ (B)

$RCH_2\overset{+}{N}Me_2CH_2CH{=}CHR'$

$\downarrow$

$R\bar{C}H\overset{+}{N}Me_2CH_2CH{=}CHR'$

$\nearrow$ $RCH(NMe_2)CH_2CH{=}CHR'$ (C)

$\searrow$ $RCH(NMe_2)CHR'CH{=}CH_2$ (D)

On orbital-symmetry considerations, A and C are forbidden as concerted processes. As non-concerted processes (e.g., the radical-in-cage mechanism), they incur a higher energy of activation, but do not share the disadvantage of B and D as to entropy in that several degrees of freedom are restricted in assuming the transition-state conformation.

The Sommelet change can be regarded as a special case of D, slightly more favourable on grounds of entropy but requiring the sacrifice of benzene resonance at the critical stage.

The particular case of migration, with inversion, of an allyl group to the γ-position of a second allyl is forbidden.

The courses A and B are exemplified by the behaviour of the ion (27) (p. 102). Ollis showed that a variety of allylphenacyl ions e.g., $PhCOCH_2\overset{+}{N}Me_2CH_2CH{=}CMe_2$, followed exclusively course D in rearrangement [32]. He further observed that a propargylammonium ion can rearrange with analogous 'inversion' to allenyl. Baldwin had previously recorded the sulphonium rearrangement:

$$(PhC{\equiv}CCH_2)_2\overset{+}{S}Et \longrightarrow CH_2{=}C{=}CPhCHSEt \quad (\text{with } C{\equiv}CPh \text{ on CH})$$

Schöllkopf has observed a similar change in a fluorenyl propargyl ether. In contrast, Morris found [31] no inversion of allyl in the rearrangement of $MeCO\bar{N}\overset{+}{N}Me_2CH_2CH{=}CMe_2$, in which the radical-in-cage mechanism was indicated by C.I.D.N.P. Some systems of the type: $RCO\bar{N}\overset{+}{\dot{N}}R'_2 \longleftrightarrow RCO\dot{N}NR'_2$ are fairly stable free radicals of long life.

In the rearrangement of fluorenyl ethers, Schöllkopf found allylic inversion:

9-H, 9-O(RCHCH=CHR')-fluorene ⟶ 9-HO, 9-(R'CHCH=CHR)-fluorene

(R = H, R′ = Me and R = Me, R′ = H)

Much earlier, Cope had observed a similar inversion in the rearrangement of (crotyl)methylaniline oxide, $O{-}\overset{+}{N}MePhCH_2CH{=}CHMe$.

The production of sulphonium ylides and their rearrangement attract special interest chemically, and have also been assigned great biological importance. In contrast with nitrogen, sulphur has d-electrons which can take part in an expansion of the octet. In particular, a sulphonium ylide can be formulated alternatively as an alkylidene-sulphurane:

$$R_2\overset{+}{S}-\overset{-}{C}R'_2 \longleftrightarrow R_2S{=}CR'_2$$

If R′ is an electron-attracting group, the sulphonium ylide, in contrast with the ammonium analogue, is isolable and may be quite stable [33]. Accordingly, the related sulphonium ion loses a proton very easily, even to aqueous carbonate. Unlike ammonium ions, then, some sulphonium ions can undergo rearrangement 'under physiological conditions'.

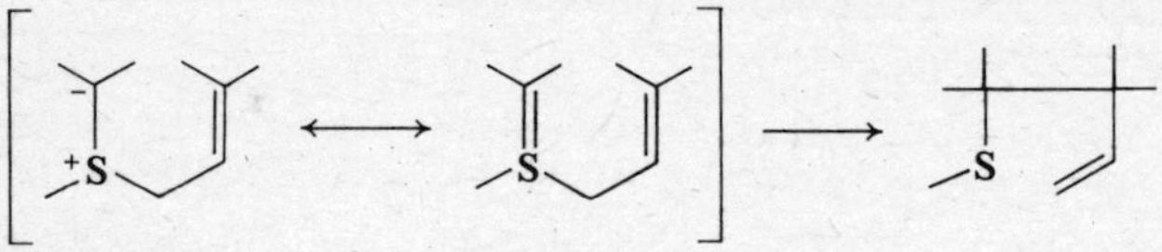

The migration of an allyl group with inversion can be formulated [34] as a sigmatropic (five-centre) reaction impelled by the change in valency of sulphur from four to two. This formulation also accommodates the Sommelet-type change sometimes encountered in sulphonium ions; thus tribenzylsulphonium gives *o*-methylbenzhydryl benzyl sulphide, *o*-$MeC_6H_4CHPhSCH_2Ph$. These alternative possibilities of a concerted sigmatropic route and a radical-in-cage path are manifested in a wide range of formally electrophilic rearrangements [33].

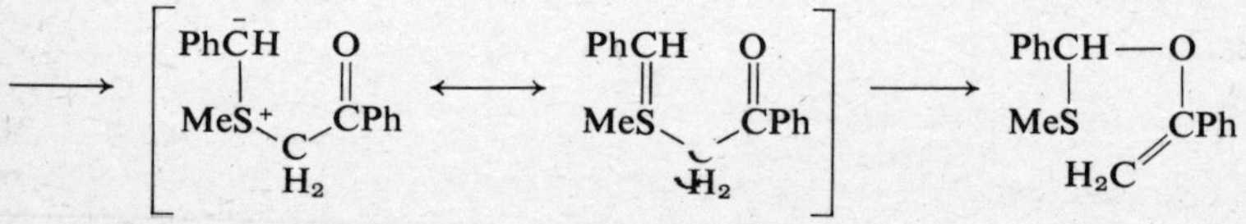

The tendency to follow the cyclic sigmatropic route is well shown by the behaviour of the ylide $PhCO\overset{-}{C}H\overset{+}{S}MeCH_2Ph$ (37). Under some conditions it yields the product of a direct 1,2-shift, i.e., $PhCOCH(SMe)CH_2Ph$ (Böhme), but the usual product in this and related cases is that of the cyclic process (Ruiz; Ratts; Schöllkopf [34]), which involves, in the first step, prototropic change to an isomeric ylide. The curious structure of the product—a half acetal, half mercaptal, derivable from benzaldehyde + enolic acetophenone and methanethiol—is well established. The rearrangement (*vide infra*) of sulphoxides and of sulphilimines can be formulated similarly. A more remote parallel is a rearrangement of sulphoxides first noted by Pummerer in 1910:

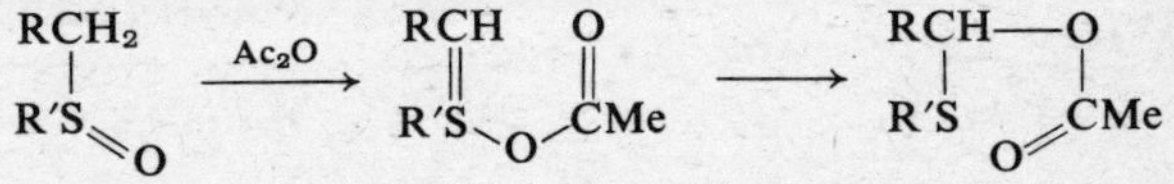

$$\begin{array}{c} RCH_2 \\ | \\ R'S{=}O \end{array} \xrightarrow{Ac_2O} \begin{array}{c} RCH \quad O \\ \| \quad\;\; \| \\ R'S{-}O{-}CMe \end{array} \longrightarrow \begin{array}{c} RCH{-}O \\ | \qquad | \\ R'S \quad O{=}CMe \end{array}$$

It appeared (1964–66) that squalene (38), a 'Charing Cross' in biosynthetic routes to triterpenoids and steroids, arises from two molecules of farnesol (39), in one of which C_1 retains its configuration (disclosed by isotopic labelling) with acceptance of one hydrogen atom from a reducing co-enzyme, while in the other molecule C_1 is inverted.

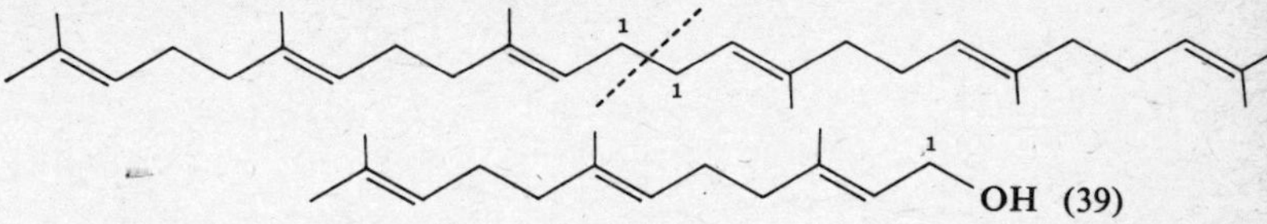

1 1 (38)

1 OH (39)

It was surmised that this unsymmetrical mode of assemblage of identical C_{15} units might be achieved by a 1,2-sulphonium ion rearrangement, something like Scheme 7.

Scheme 7

$$2RCH_2OH + (enzyme)SH \longrightarrow (RCH_2)_2\overset{+}{S}(enzyme) \xrightarrow{base} \begin{array}{c} RCHS(enzyme) \\ | \\ RCH_2 \end{array} \xrightarrow{2H} RCH_2CH_2R + (enzyme)SH$$

$(R = C_{14}H_{23})$

Model experiments followed in which diallyl sulphides were converted into diallyl-methyl- or -ethyl-sulphonium ions by treatment with trimethyl- or triethyl-oxonium borofluoride, and then rearranged with alkali. Alternatively, the sulphides were treated with benzyne. In each case the resulting methyl- or phenyl-thiol was submitted to hydrogenolysis, e.g., by lithium in liquid ammonia. The use of diallylphenyl-sulphonium ions, or of the combination: diallyl sulphide-benzyne, tended to give some of the 'uninverted' product of the radical-in-cage mechanism; the method using diallyl-ethylsulphonium was preferred. Thus Baldwin found that geranyl chloride with sodium sulphide gave the 'normal' product, digeranyl sulphide. The derived ethylsulphonium salt rearranges with inversion (n.m.r.) and hydrogenolysis then gave the C_{20} hydrocarbon

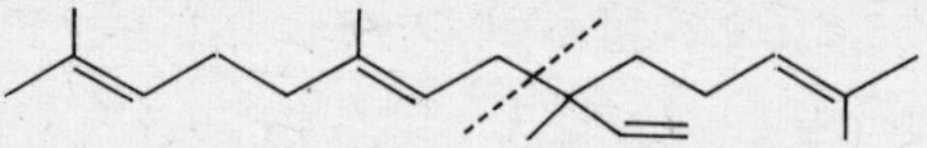

Ollis synthesised squalene from farnesol [35]. Farnesyl chloride with thiourea gave the thiuronium chloride which, by hydrolysis and atmospheric oxidation, furnished difarnesyl disulphide. Triphenylphosphine extracted one sulphur atom, giving farnesyl nerolidyl sulphide (40); the constitution, involving an allylic inversion, was confirmed by n.m.r. Rearrangement by combination with benzyne gave 12-phenylthiosqualene (41) with a compensating inversion of the nerolidyl moiety, and hydrogenolysis finally yielded squalene.

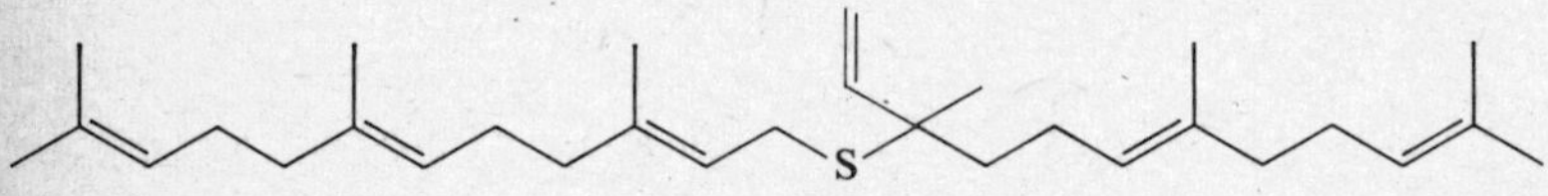

S

(40)

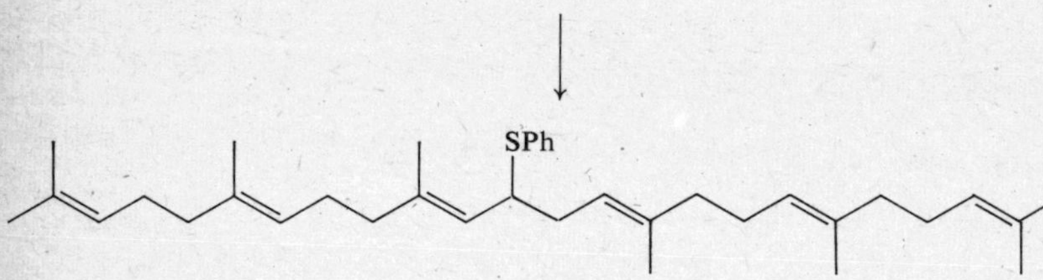

SPh

(41)

Li/NH_3

Squalene

In a variant [36] of the Baldwin–Ollis synthetic scheme, diallyl sulphides themselves can suffer sigmatropic rearrangement:

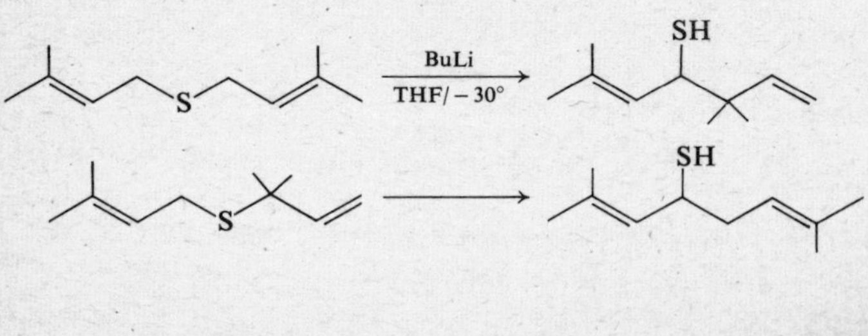

S —BuLi, THF/ − 30°→ SH

S ⟶ SH

Baldwin applied [37] a related ether rearrangement to the synthesis of the anomalous monoterpene artemisia alcohol:

O — BuLi / THF → OH

Boekelheide used 1,2-migrations involving contraction of rather large rings (e.g. 42) to achieve interesting syntheses [38]. The first methylation was effected by the salt $(MeO)_2\overset{+}{C}H\ \bar{B}F_4$. A second sequence, commencing with reaction between the dibromide (43) and sodium sulphide, gave the cyclotetradecaheptaene derivative (44).

$HSCH_2$ … CH_2SH + $BrCH_2$ … CH_2Br → S … S → MeS^+ … ^+SMe (42)

(42) —NaH→ [mixture] —1. Methylation, 2. t-BuOK→

$BrCH_2$ … CH_2Br, Me (43) → Me, Me (44)

The sulphur analogue of the Meisenheimer rearrangement would be the transformation of a sulphoxide into a sulphenic ester. When allyl *p*-tolyl sulphoxide was heated with piperidine it gave toluene–sulphenpiperidide:

$$\left[\begin{array}{c} O{=}SC_7H_7 \\ | \\ CH_2 \\ | \\ CH{=}CH_2 \end{array} \longleftrightarrow \begin{array}{c} \bar{O}{-}\overset{+}{S}C_7H_7 \\ | \\ CH_2 \\ | \\ CH{=}CH_2 \end{array}\right] \longrightarrow \begin{array}{c} OSC_7H_7 \\ | \\ CH_2 \\ | \\ CH{=}CH_2 \end{array} \xrightarrow{C_5H_{11}N} \begin{array}{c} C_5H_{11}NSC_7H_7 \\ + \\ H_2C{=}CHCH_2OH \end{array}$$

Sulphenic esters, in fact, undergo aminolysis by piperidine.

It appears [39], however, that the sulphoxide greatly predominates at equilibrium (contrast the position in the Meisenheimer change). This is attributable to the strength of the SO link of a sulphoxide which has a high degree of double-bond character. Lithium benzyloxide reacts with a sulphenyl chloride, ArSCl, giving the sulphenic ester $ArSOCH_2Ph$, which rearranges, apparently completely, to aryl benzyl sulphoxide at 100°. When a lithium allyloxide is used, the sulphenic ester initially formed rapidly rearranges to sulphoxide with inversion of the allyl group.

Sulphoxides, RR′SO, are chiral (pyramidal disposition of bonds to sulphur) and can be racemised by reversible isomerisation to the achiral sulphenate. Suggestively, allyl sulphoxides racemise at about 40° (sigmatropic equilibration); benzyl sulphoxides at about 140° (radical-in-cage equilibration); other sulphoxides at about 200° (direct thermal racemisation).

The lithium salt of optically active methallyl alcohol gave with sulphenyl chloride an active sulphenic ester (chiral carbon atom) which changed into active crotyl sulphoxide (chiral sulphur atom):

$$CH_2{=}CHCH(Me)OSAr \longrightarrow MeCH{=}CHCH_2S^+(O^-)Ar$$

A similar transfer of chirality (from N to C) has been observed [40] in the active quaternary ammonium ion (45) and in the active amine oxide (46) (Inouye). These varied observations emphasise the concerted character of the rearrangements concerned.

$$PhCH_2{-}\overset{\overset{\displaystyle Ph}{|}}{\underset{\underset{\displaystyle Me}{|}}{N^+}}{-}CH_2CH{=}CH_2 \xrightarrow[DMSO]{BuOK} PhCH_2\overset{\overset{\displaystyle PhNMe}{|}}{C}HCH{=}CH_2 \quad (15\%)$$

(45)

$$Ar{-}\overset{\overset{\displaystyle Et}{|}}{\underset{\underset{\displaystyle O^-}{|}}{N^+}}{-}CH_2CH{=}CHMe \longrightarrow Ar\overset{\overset{\displaystyle Et}{|}}{N}OCHMeCH{=}CH_2$$

(46)

The relation between sulphinic esters and sulphones, on the other hand, shows the general tendency for quadrivalent sulphur to become sexivalent. Many sulphinic esters change when heated into sulphones (conceivably a 1,2-electrophilic migration to sulphur):

$$\mathrm{R{-}O{-}\overset{+}{S}(O^-){-}R'} \longrightarrow \mathrm{\bar{O}{-}\overset{2+}{S}(O^-)(R){-}R'}$$

The reaction, however, is significantly favoured by ionising solvents, and mixture experiments indicate [41] that the process is intermolecular:

$$\mathrm{ArS(=O)OCHPh_2} \longrightarrow \mathrm{ArS(=O){-}O^-} + \overset{+}{\mathrm{C}}\mathrm{HPh_2} \longrightarrow \mathrm{Ar\overset{+}{S}(=O)(CHPh_2){-}\bar{O}}$$

On the other hand, Stirling reports [42] a notable example of an evidently concerted rearrangement of an optically active sulphinic ester:

$$\mathrm{ArSOOCHMeC{\equiv}CH} \longrightarrow \mathrm{ArSO_2CH{=}C{=}CHMe}$$

The optically active product owes its asymmetry to the presence of an allene system.

The rearrangement of sulphilimines to derivatives of thiohydroxylamine (Challenger) parallels that of sulphoxides:

$$\mathrm{PhCH_2{-}S{-}CH_2CH{=}CH_2} + \text{chloramine-T} \longrightarrow \left[\mathrm{PhCH_2(H_2C{=}CHCH_2)\overset{+}{S}{-}\overset{-}{N}Ts} \longleftrightarrow \mathrm{PhCH_2(H_2C{=}CHCH_2)S{=}NTs}\right] \xrightarrow{\text{heat}} \mathrm{PhCH_2{-}S{-}NTs(CH_2CH{=}CH_2)}$$

A change from 4- to 6-valent sulphur, with radical migration, was observed long ago in the conversion of sodium alkyl sulphites into alkanesulphonates (i.e., $\mathrm{RO{-}SO{-}O^- \rightarrow R{-}SO_2{-}O^-}$) which is promoted by small amounts of such salts as potassium iodide, suggesting the sequence:

$$\mathrm{ROSO_2^-} + \mathrm{I^-} \longrightarrow \mathrm{RI} + \mathrm{SO_3^{2-}}$$

$$\mathrm{ROSO_2^-} + \mathrm{RI} \longrightarrow \mathrm{ROS(R)O_2} \xrightarrow{\mathrm{H\bar{O}}} \mathrm{\bar{O}S(R)O_2}$$

Dialkyl sulphites are effective alkylating agents. When treated with alkali or tertiary amines, they are transformed into alkyl alkanesulphonates. Apparently the alkaline reagent produces some of the anion, $ROSO_2^-$; this is alkylated on sulphur by a second molecule of ester, giving alkyl alkanesulphonate and regenerating the anion [43].

Migration from oxygen to phosphorus. Such processes may be considered here. For the **Arbusov rearrangement** [44] of phosphorus esters in presence of alkyl halides

$$(EtO)_3P \xrightarrow{EtI} EtPO(OEt)_2$$

a plausible explanation is suggested by the fact that, when methyl iodide is used as promoter, the methyl group becomes linked to phosphorus:

$$(EtO)_3P + MeI \longrightarrow (EtO)_3\overset{+}{P}Me\ I^- \longrightarrow EtI + (EtO)_2PMeO$$

Savage, and Tieckelmann, showed that compounds, $X_2P(OR)$, can rearrange when heated alone, giving $X_2P(R){=}O$, with inversion when R = allyl. The group X can be Ph or OEt. The generality of this type of reaction has been emphasised by the work of Razumova.

References

1. T. S. Stevens *et al.*, *J. Chem. Soc.*, 1928, 3193; 1932, 55, 1926.
2. R. A. W. Johnstone and T. S. Stevens, *J. Chem. Soc.*, 1955, 4487.
3. G. Wittig *et al.*, *Annalen*, 1951, **572**, 1.
4. S. W. Kantor and C. R. Hauser, *J. Amer. Chem. Soc.*, 1951, **73**, 4122.
5. M. Sommelet, *Compt. rend. Acad. Sci. Paris*, 1937, **205**, 56.
6. C. R. Hauser *et al.*, *J. Amer. Chem. Soc.*, 1957, **79**, 5512; *J. Org. Chem.*, 1961, **26**, 2979.
7. J. Casanova, N. D. Werner, and R. E. Schuster, *J. Org. Chem.*, 1966, **31**, 3473; S.-I. Yamada, M. Shibasaki, and S. Terashima, *Chem. Comm.*, 1971, 1008.
8. G. Wittig *et al.*, *Annalen*, 1942, **550**, 260; 1947, **557**, 205; 1949, **562**, 192; 1954, **588**, 145; 1957, **605**, 69.
9. D. Y. Curtin and W. R. Proops, *J. Amer. Chem. Soc.*, 1954, **76**, 494.
10. U. Schöllkopf, H. Schäfer, and D. Walter, *Tetrahedron Letters*, 1968, 2809.
11. R. West, R. Lowe, H. F. Stewart, and A. Wright, *J. Amer. Chem. Soc.*, 1971, **93**, 282.

12. W. F. Cockburn, R. A. W. Johnstone, and T. S. Stevens, *J. Chem. Soc.*, 1960, 3340, 3346.
13. E. Grovenstein and G. Wentworth, *J. Amer. Chem. Soc.*, 1957, **79**, 4985; 1967, **89**, 1852, 2348.
14. H. E. Zimmerman *et al.*, *J. Amer. Chem. Soc.*, 1957, **79**, 5455; 1961, **83**, 1196.
15. S. Wawzonek and E. Yeakey, *J. Amer. Chem. Soc.*, 1960, **82**, 5718.
16. R. West *et al.*, *J. Amer. Chem. Soc.*, 1967, **89**, 5050; 1969, **91**, 5184.
17. F. A. Carey and L. J. Hayes, *J. Amer. Chem. Soc.*, 1970, **92**, 7613.
18. J. Meisenheimer, *Ber.*, 1919, **52**, 1667.
19. A. H. Wragg, T. S. Stevens, and D. M. Ostle, *J. Chem. Soc.*, 1958, 4057; N. Castagnoli *et al.*, *Tetrahedron*, 1970, **26**, 4319.
20. U. Schöllkopf *et al.*, *Annalen*, 1967, **703**, 77.
21. L. D. Quinn and F. A. Shelburne, *J. Org. Chem.*, 1965, **30**, 3135.
22. E. J. Grubbs *et al.*, *J. Amer. Chem. Soc.*, 1967, **89**, 2234; 1969, **91**, 2022.
23. W. R. Bamford and T. S. Stevens, *J. Chem. Soc.*, 1952, 4675; W. Ando *et al.*, *J. Amer. Chem. Soc.*, 1969, **91**, 5164, 6516.
24. H. Hellmann *et al.*, *Annalen*, 1960, **631**, 82; 1963, **662**, 188.
25. J. H. Brewster and M. W. Kline, *J. Amer. Chem. Soc.*, 1952, **74**, 5179.
26. B. J. Millard and T. S. Stevens, *J. Chem. Soc.*, 1963, 3397.
27. U. Schöllkopf *et al.*, *Annalen*, 1961, **642**, 1; 1965, **683**, 42.
28. E. F. Jenny *et al.*, *Angew. Chem. Internat. Edn.*, 1962, **1**, 155; 1965, **4**, 441.
29. U. Schöllkopf *et al.*, *Tetrahedron Letters*, 1969, 2619, 3415.
30. H. Fischer and J. Bargon, *Accounts of Chem. Research*, 1969, **2**, 110.
31. D. G. Morris, *Chem. Comm.*, 1969, 1345.
32. R. W. Jemison and W. D. Ollis, *Chem. Comm.*, 1969, 294.
33. J. E. Baldwin *et al.*, *Chem. Comm.*, 1968, 537, 538; 1970, 576; *J. Amer. Chem. Soc.*, 1969, **91**, 3646.
34. U. Schöllkopf, J. Schossig, and G. Ostermann, *Annalen*, 1970, **737**, 158.
35. W. D. Ollis *et al.*, *Chem. Comm.*, 1969, 99.
36. V. Rautenstrauch, *Helv. Chim. Acta*, 1971, **54**, 739.
37. J. E. Baldwin, J. DeBernardis, and J. E. Patrick, *Tetrahedron Letters*, 1970, 353.
38. V. Boekelheide *et al.*, *Tetrahedron Letters*, 1970, 1197, 1207.
39. K. Mislow *et al.*, *J. Amer. Chem. Soc.*, 1966, **88**, 3139; 1968, **90**, 4869.
40. R. K. Hill and T. H. Chan, *J. Amer. Chem. Soc.*, 1966, **88**, 866.
41. A. H. Wragg, J. S. McFadyen, and T. S. Stevens, *J. Chem. Soc.*, 1958, 3603.
42. G. Smith and C. J. M. Stirling, *J. Chem. Soc.* (C), 1971, 1530.
43. H. F. van Woorden, *Chem. Rev.*, 1963, **63**, 562.
44. G. M. Kosolapoff, *Org. Reactions*, 1951, **6**, 276.

Review Articles

Valuable background information as well as a treatment of the various types of rearrangement is to be found in:

D. J. Cram, *Fundamentals of Carbanion Chemistry*, Academic Press (New York), 1965.

U. Schöllkopf, *Angew. Chem. Internat. Edn.*, 1970, **9**, 763.

Stevens and Wittig rearrangements

H. E. Zimmerman in *Molecular Rearrangements* Part 1, pp. 345–406, ed. de Mayo, Interscience (New York), 1963.

Wittig rearrangement

D. L. Dalrymple *et al.* in *Chemistry of the Ether Linkage*, pp. 618–628, ed. Patai, Interscience (New York), 1967.

Fritsch rearrangement

T. L. Jacobs, *Org. Reactions*, 1949, **5**, 40.

Stevens and Sommelet–Hauser rearrangements

S. H. Pine, *Org. Reactions*, 1970, **18**, 403.

6

Further Alkali-Induced Rearrangements

This chapter deals with several rather loosely connected topics:

(a) rearrangements, evidently to be classified as electrophilic, in which migration takes place over a greater range than a 1,2-shift;

(b) processes where an anionic reagent adds to a carbonyl group to give an anion in which a 1,2-migration follows;

(c) alkali-induced 1,2-rearrangements which appear to involve a three-membered ring intermediate;

(d) reactions in which a 1,3-shift takes place to a doubly-linked destination atom bearing a lone pair of electrons. It may be only formally possible to associate such reactions with electrophilic rearrangements.

Longer-range Electrophilic Migrations

In contrast to 1,2-electrophilic migrations, a 1,3-shift is rarely encountered. This is easily understood; a case where a 1,3-shift is favourable is usually still more disposed towards an elimination reaction, e.g., a Hofmann elimination in the following sequence where X = Y = carbon, Z = ammonium nitrogen.

$$\text{R}\text{—X—Y—}\ddot{\text{Z}} \;\not\longleftarrow\; \text{—}\ddot{\text{X}}\text{—Y—Z(R)—} \longrightarrow \text{X=Y} + \text{:Z(R)—}$$

One example (where elimination would produce a ketene), which offers a useful synthetic route to γ,δ-unsaturated carboxylic acids, is the action of a strong base on an allyl ester [1]:

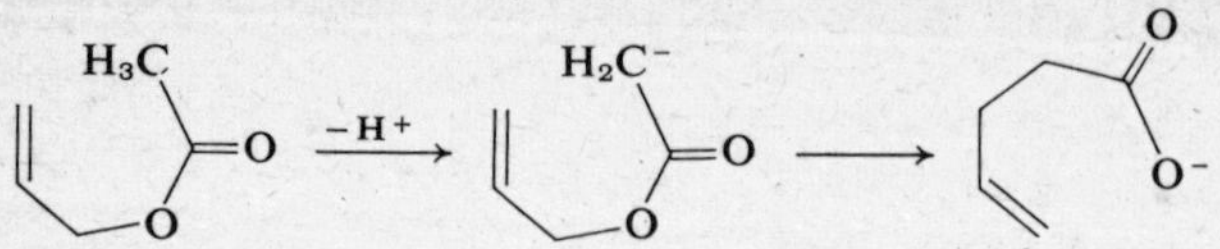

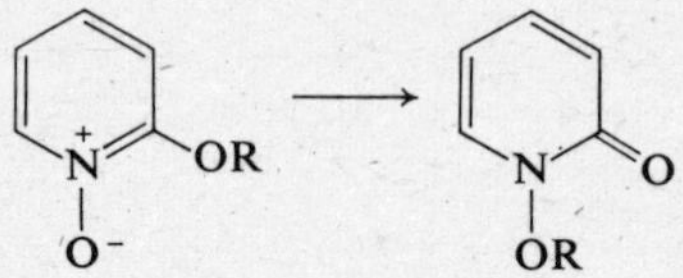

(1) (2)

On the other hand, longer-range shifts are not unusual. Many are classifiable as intramolecular S_N2 processes, not symmetry-forbidden, and of these the most extensive studies are due to Smiles and Truce. In a rather special case of O → O migration (1) → (2), the change is intramolecular and little affected by polar substituents. Migration of a chiral group (R = PhCHD) proceeds with retention of configuration at carbon. No radical intermediates are detected by C.I.D.N.P., and the process is described [2] as a [1,4] suprafacial sigmatropic shift (see p. 171).

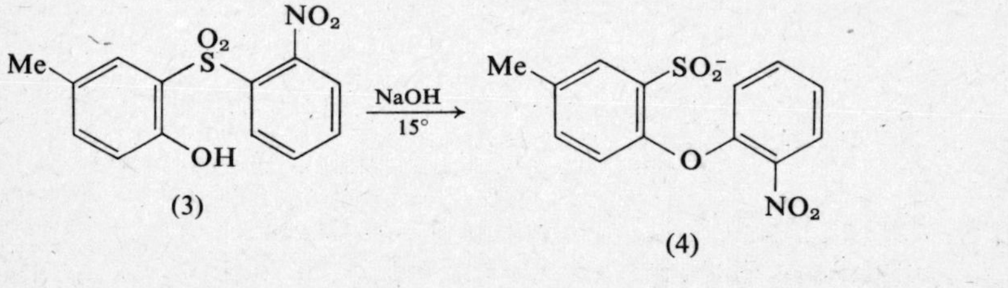

(3) (4)

Smiles rearrangement. In 1931, Smiles first reported [3] a type of rearrangement of which the transformation (3) → (4) is typical. This reaction is very naturally regarded as involving an intramolecular nucleophilic substitution on the nitrophenyl nucleus by phenoxide anion and the process may be given the general expression of Scheme 1. The essential conditions for the reaction may be summarised as follows:

(a) the YH group must be sufficiently acidic to be deprotonated by HO^-, giving nevertheless a sufficiently basic anion;

(b) X^- must be a suitable leaving-group;

(c) the aromatic nucleus in which nucleophilic substitution occurs (i.e., ring B; see Scheme 1) must usually carry electron-attracting substituents.

Scheme 1

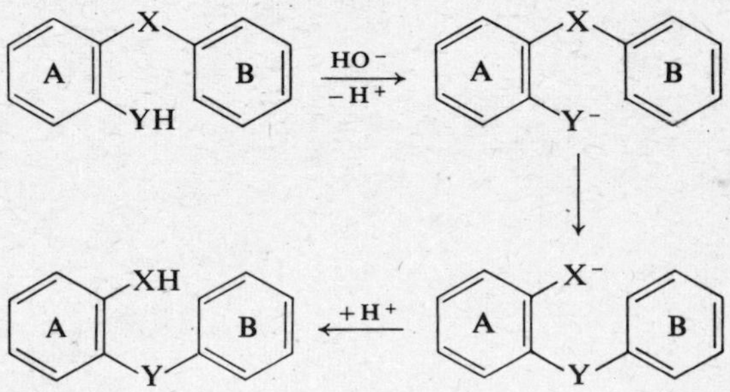

For —X—, the groups $—SO_2—$, —SO—, and —S— form a series in order of diminishing effectiveness; cf. sulphinic acids ($ArSO_2H$) are stronger acids than thiophenols (ArSH) by some five pK units. The reaction frequently succeeds when X = O but not, understandably, when X = NH. While the transformation was first observed in a case in which an unsubstituted naphthyl group served as ring B, a phenyl group usually requires one of the substituents, *o*-NO_2, *p*-NO_2, or *p*-SO_2R, diminishing in effectiveness in that order.

In a series of experiments with X = SO_2 and YH = OH, ring B carried a nitro group and a further substituent. With an *o*-NO_2 group, further *p*-substituents facilitated the reaction in the order NO_2 > PhCO > CO_2^- > Cl > H; with a *p*-NO_2 group, further *o*-substituents were favourable in the order NO_2 > PhCO > Cl > H.

When YH = OH, the situation is favourable; the anion (Y^-) is easily produced but still sufficiently basic. The presence of a nitro group *p*- to the hydroxyl group in ring A retards the rearrangement, no doubt by reducing the basicity of O^-. With YH = RNH, the situation is more complex. For R = H or alkyl, extraction of a proton is very difficult and rearrangement is hard to effect; when R = acyl, the situation is quite favourable, but not when R = sulphonyl (Y^- now deficient in basicity). In the case of the ether (5) with a very favourably substituted ring B and YH = NH_2, rearrangement takes place in the absence of alkali, the uncharged NH_2 group being sufficiently basic.

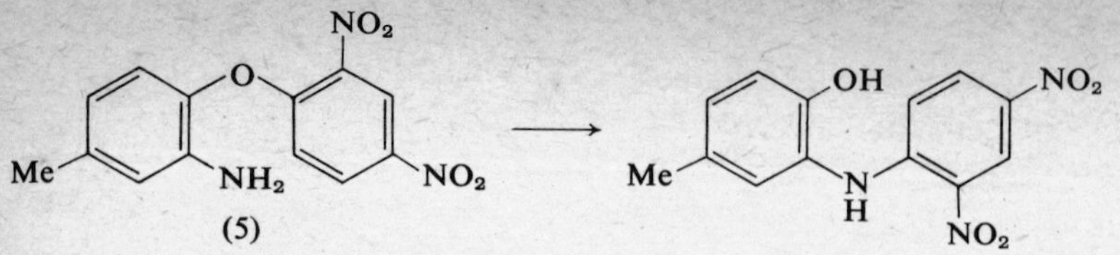

Analogous 1,5-rearrangements have been recorded, for example where YH = CONHR:

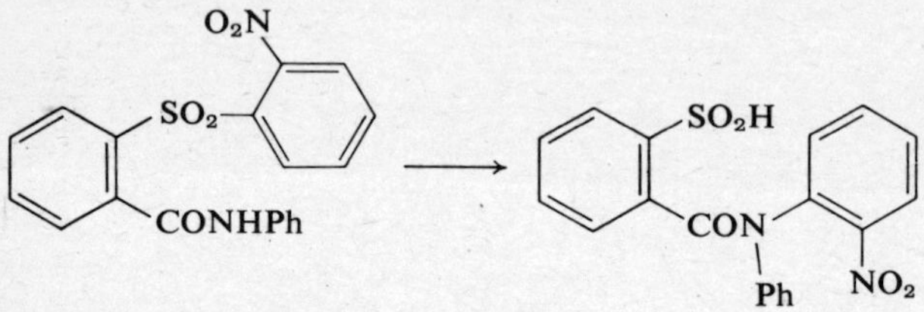

Foreseeably, the reaction fails where YH = CO_2H.

Compounds in which ring A (see Scheme 1) is replaced by two methylene groups ($HYCH_2CH_2XAr$) can undergo the Smiles rearrangement, but less readily, presumably because the more flexible system will be found less often in the conformation required for reaction. Conversely, the rearrangement of the diaryl substrates can be accelerated spectacularly by substitution *o*- to X in ring A [4]. In the unsubstituted material, the predominant conformation of the molecule will be (6), unfavourable for rearrangement, but a bulky *o*-substituent will tend to impose the conformation (7), suitable for a displacement reaction on ring B.

(6) (7)

The rearrangement can be effected in heterocyclic systems. Thus, the compound (8; R = 2- or 4-NO_2) rearranges in alkaline or in acidic solution. In an acidic medium, any protonation of the ring B pyridine nitrogen atom (not, however, the most basic centre in the molecule!) will greatly enhance the rate of attack by the unprotonated NH_2 group. The electron-attractive tetrazole system, as in (9), can also migrate.

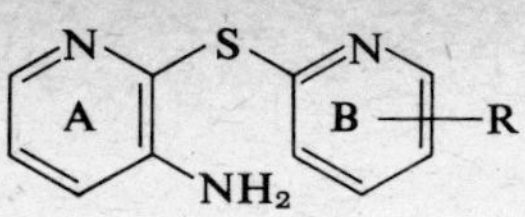

(8)

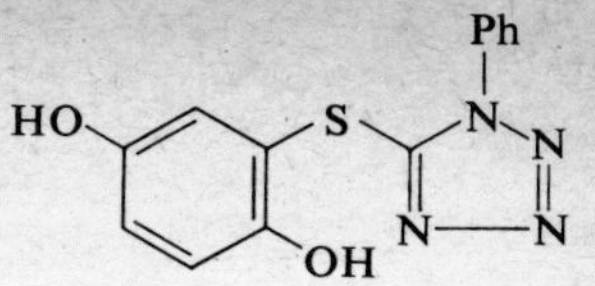

(9)

Under acid conditions, the rearrangement of a hydroxy-sulphone (10) may be reversed.

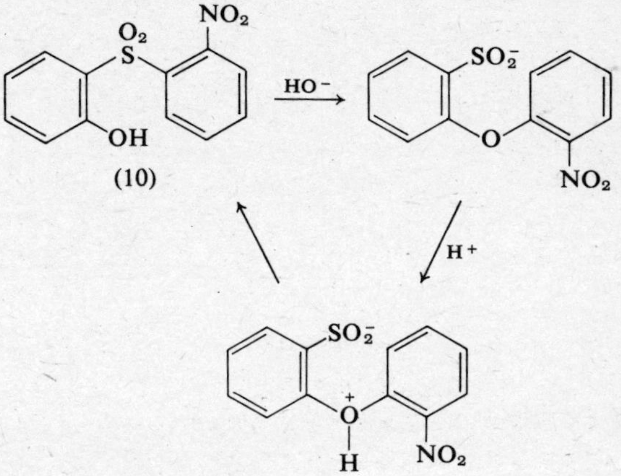

(10)

Two successive Smiles rearrangements are concerned (see Scheme 2) in the instructive transformation and regeneration of the hydroxy-sulphone (11).

Scheme 2

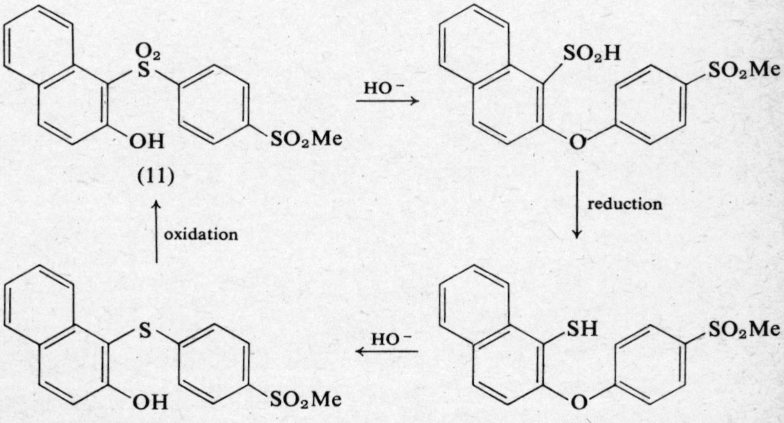

(11)

In conclusion, two less typical cases may be cited in which X and Y are connected by an aliphatic chain (Schemes 3 [5] and 4). In Scheme 4, note that the instability of the *N*-sulphinic acid allows a second Smiles rearrangement.

Scheme 3

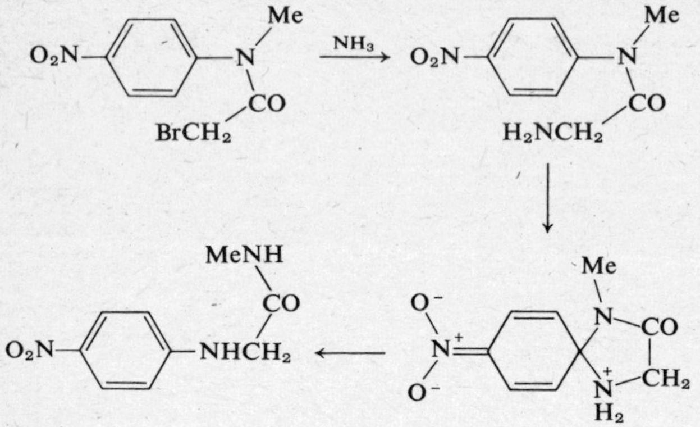

Scheme 4

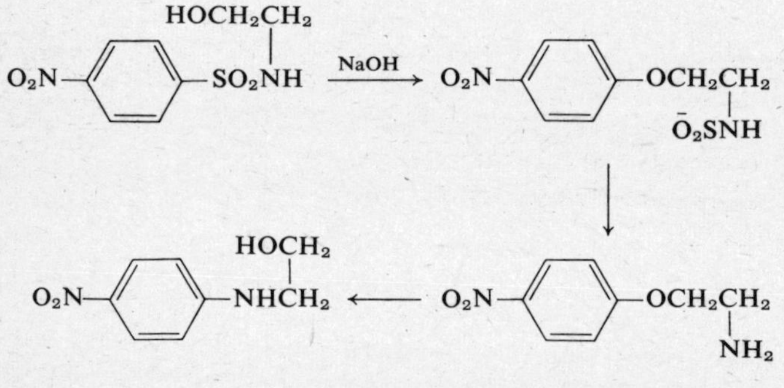

Truce rearrangement. In 1958 Truce observed [6] that, when the intensely reactive nucleophile $ArCH_2^-$ takes the place of the ArO^- nucleophile of a typical Smiles reaction, rearrangement of a sulphone (12) proceeds readily in the absence of promoting nitro substituents in the migrating nucleus.

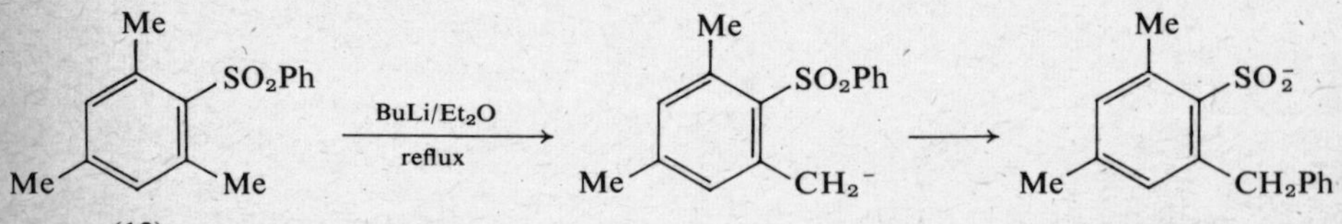

The metallation stage is rapid, not rate-determining. The process as a whole is affected by substitution in the migrating nucleus in a way not clearly understood, and it is strongly subject to steric influences, somewhat in the same way as the Smiles rearrangement.

The reactions of naphthalene derivatives show some peculiarities. Thus, migration *to* a *peri*-methyl carbon occurs normally but the migration *of* a 1- or 2-naphthyl group takes an unexpected course, as with the sulphone (13). Similarly, the 1-naphthyl sulphone gives a (2-naphthyl)phenylmethane derivative.

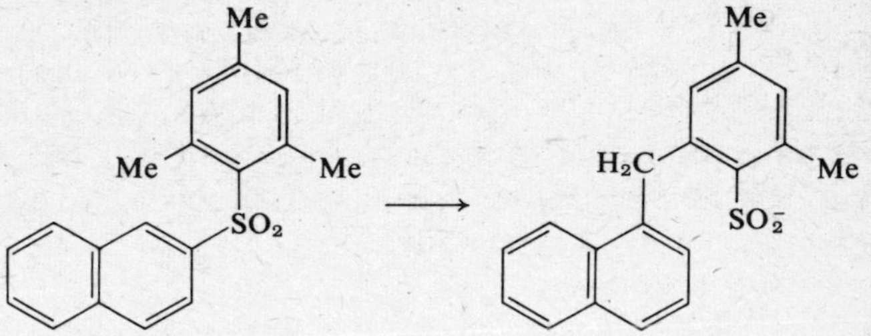

Me Me Me SO_2 → Me H_2C Me SO_2^-

(13)

Truce [7] and Drozd have shown independently that a dihydrothioxanthene-*S*-dioxide (14) can be produced in these reactions, and Truce proposes the mechanism shown in Scheme 5.

Scheme 5

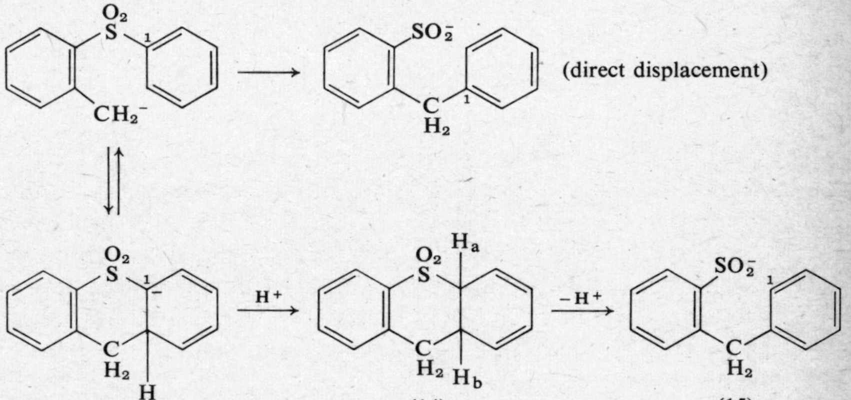

O_2 S 1 CH_2^- → SO_2^- C 1 H_2 (direct displacement)

⇅

O_2 S 1 C H_2 H $\xrightarrow{H^+}$ H_a O_2 S C H_2 H_b $\xrightarrow{-H^+}$ SO_2^- 1 C H_2

(14) (15)

When the migrating ring is part of a naphthalene system, the equilibrium between the benzyl and cyclohexadienyl anions will be displaced strongly in favour of the latter, since less resonance energy is sacrificed than in the case of a migrating unfused benzene ring. In ether solution, no proton is available to form compound (14), and the reaction is limited to a 'normal' direct displacement. On the other hand, dimethyl sulphoxide as solvent can supply a proton with possible formation of (14), and thence the 'abnormal' product (15). When (14), separately prepared, is introduced into a strongly basic medium (BuLi/Et_2O or t-BuOK/DMSO), the acidic proton (H_a) is removed and equilibrium is re-established so that (14) gives the same product as the original sulphone under the same conditions. With a weaker base (EtONa/EtOH), however, only the concerted process is possible, i.e., attack on H_b with separation of SO_2^-, thereby restoring the aromaticity of the migrating ring through an elimination reaction and yielding the product (15). When the migrating ring bears a suitable 'marking' substituent, the 'abnormal' product (15) is distinct from the product of direct displacement.

Alkali-induced 1,2-Migrations in Ketones

The **benzilic acid rearrangement** (16) → (17), recorded by Liebig in 1838, differs notably from the alkali-induced 1,2-migrations described in Chapter 5, in that the alkali does not abstract a proton from the substrate but adds a basic anion. Moreover, the migrating-group (often aryl) moves to an electron-poor carbonyl carbon, associating the change with the class of nucleophilic rearrangements.

$$\underset{(16)}{\begin{matrix}\text{Ph—C=O}\\ |\\ \text{Ph—C=O}\end{matrix}} \xrightarrow[100^\circ]{\text{KOH}} \text{Ph}_2\text{C(OH)CO}_2\text{K} \xrightarrow{\text{H}^+} \underset{(17)}{\text{Ph}_2\text{C(OH)CO}_2\text{H}}$$

This type of rearrangement is by no means limited to diaryl diketones. Formally, the simplest case is the conversion of glyoxal into glycollic acid, i.e., O=CHCH=O → $HOCH_2CO_2H$. [In this case, the mechanism currently accepted for the benzilic rearrangement (*vide infra*) coincides with that assigned to the Cannizzaro reaction.] The reaction is observed with aromatic-aliphatic diketones, e.g., $PhCH_2COCOPh$ → $PhCH_2CPh(OH)CO_2H$, and with some purely aliphatic diketones, e.g. (18).

$$\underset{(18)}{\begin{matrix}\text{HO}_2\text{CCH}_2\text{—C=O}\\ |\\ \text{HO}_2\text{CCH}_2\text{—C=O}\end{matrix}} \longrightarrow \underset{\text{(Citric acid)}}{(\text{HO}_2\text{CCH}_2)_2\text{C(OH)CO}_2\text{H}}$$

Cyclic 1,2-diketones can undergo the benzilic change with ring-contraction, e.g. phenanthraquinone gives 9-hydroxyfluorene-9-carboxylic acid, a preparatively useful transition from the phenanthrene to the fluorene series.

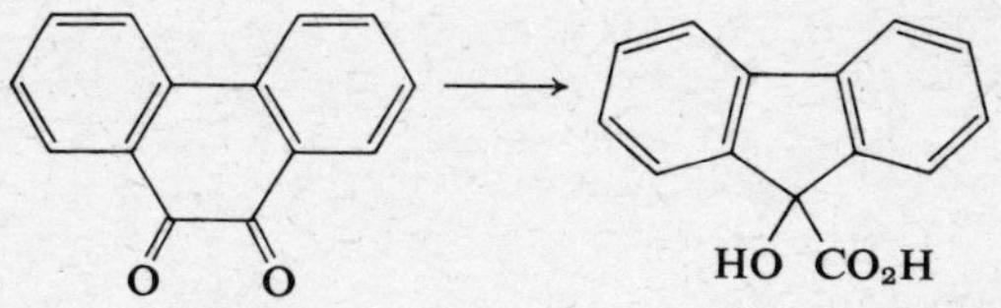

HO CO_2H

Similarly, the alicyclic cyclohexane-1,2-dione (19) gives 1-hydroxycyclopentanecarboxylic acid.

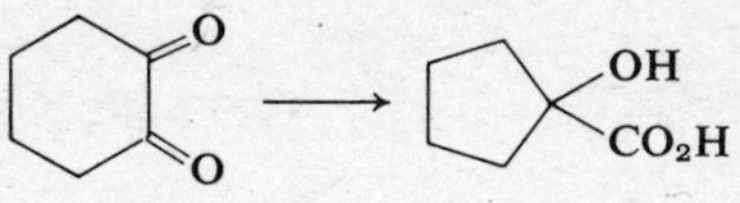

(19)

OH

CO_2H

However, the scope of the reaction in the aliphatic series is limited by competing reactions, for example:

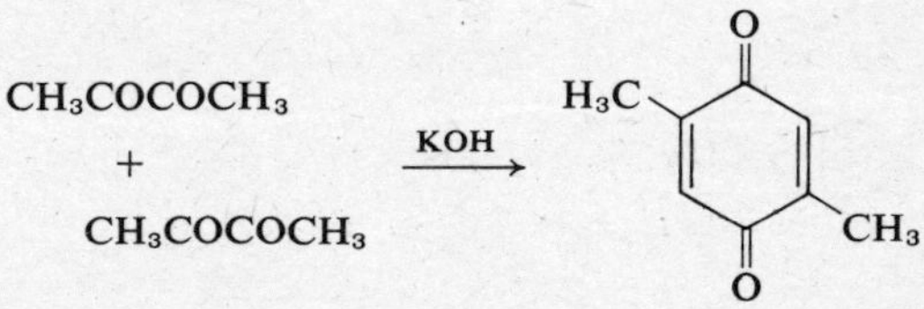

$$CH_3COCOCH_3 + CH_3COCOCH_3 \xrightarrow{KOH}$$

Evidence relating to the mechanism of the rearrangement has been provided mainly by experiments with benzil (PhCOCOPh) and substituted benzils. The reaction requires one molecule of potassium hydroxide per molecule of benzil and there is convincing evidence that benzil reacts with alkali rapidly and reversibly. Thus, a compound of stoichiometry PhCOCOPh : KOH can be isolated, and converted subsequently into potassium benzilate. Further, benzil exchanges its oxygen in alkaline ^{18}O-enriched water faster than it changes into benzilate [8].

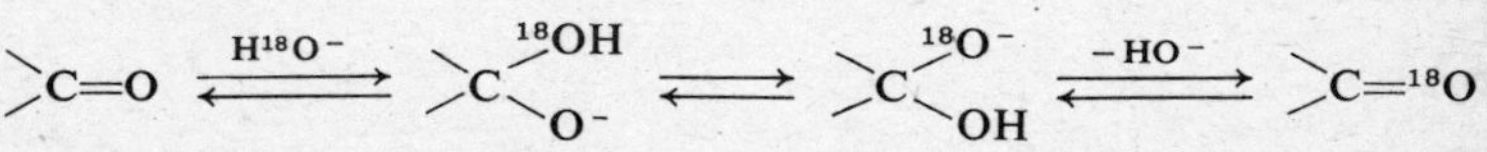

$$>C{=}O \underset{}{\overset{H^{18}O^-}{\rightleftharpoons}} >C(^{18}OH)(O^-) \rightleftharpoons >C(^{18}O^-)(OH) \underset{}{\overset{-HO^-}{\rightleftharpoons}} >C{=}^{18}O$$

The use of reagents other than alkali metal hydroxides can give rise to complicated mixtures of products, mostly from reactions unrelated to the benzilic change. Thus, with alkali metal alkoxides, reduction to benzoin is prominent. It was inferred from this that the hydrogen atom of the hydroxide ion must play an essential part in the hydroxide-promoted rearrangement.

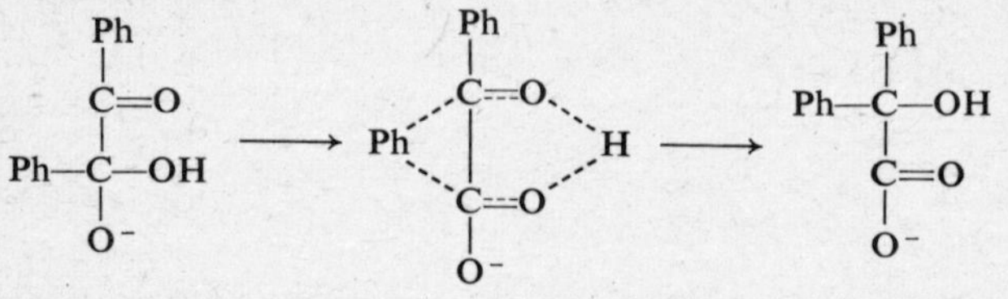

$$\mathrm{Ph{-}C(=O){-}C(Ph)(OH){-}O^-} \longrightarrow [\text{cyclic transition state: } \mathrm{Ph}, \mathrm{C{=}O}\cdots\mathrm{H}\cdots\mathrm{O{=}C{-}O^-}] \longrightarrow \mathrm{Ph_2C(OH){-}C(=O){-}O^-}$$

Later, however, Doering showed [9] that using potassium t-butoxide, most favourably but not necessarily in benzene, benzil is converted to t-butyl benzilate in high yield. Thus, transfer of hydrogen from hydroxyl is not indispensably involved in the rearrangement and, following Ingold [10], the following mechanism may be simply written:

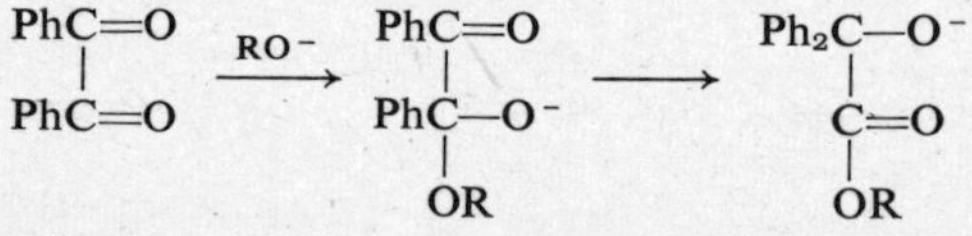

$$\mathrm{PhC{=}O{-}PhC{=}O} \xrightarrow{\mathrm{RO^-}} \mathrm{PhC{=}O{-}PhC(O^-)(OR)} \longrightarrow \mathrm{Ph_2C(O^-){-}C(=O)OR}$$

When R = t-butyl, addition of a proton gives t-butyl benzilate; when R = H, proton transfer gives benzilate anion.

The effect of substitution on the speed and direction of rearrangement has excited controversy. In symmetrically disubstituted benzils, electron-attracting substituents accelerate and electron-releasing

Scheme 6

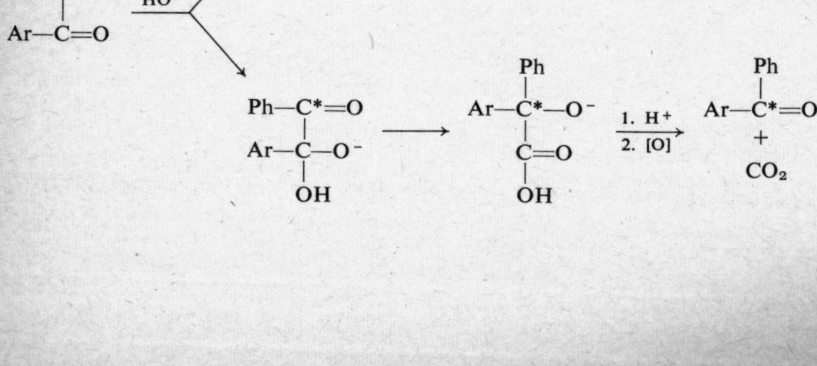

$$\mathrm{Ph{-}C^*{=}O{-}Ar{-}C{=}O} \xrightarrow{\mathrm{HO^-}} \mathrm{Ph{-}C^*(OH)(O^-){-}C(=O){-}Ar} \longrightarrow \mathrm{Ph{-}C(Ar)(O^-){-}C^*(=O)OH} \xrightarrow[2.\ [O]]{1.\ H^+} \overset{*}{\mathrm{C}}\mathrm{O_2} + \mathrm{Ph{-}C(=O){-}Ar}$$

$$\mathrm{Ph{-}C^*{=}O{-}Ar{-}C{=}O} \xrightarrow{\mathrm{HO^-}} \mathrm{Ph{-}C^*(=O){-}C(Ar)(O^-)(OH)} \longrightarrow \mathrm{Ar{-}C^*(Ph)(O^-){-}C(=O)OH} \xrightarrow[2.\ [O]]{1.\ H^+} \mathrm{Ar{-}C^*(=O){-}Ph} + \mathrm{CO_2}$$

substituents retard the rearrangement. Attempts to assign relative 'migratory aptitudes' to groups (cf. p. 23) meet the difficulty that an unsymmetrically substituted benzil gives the same benzilic acid whichever group migrates [i.e., ArCOCOAr′ → ArAr′C(OH)CO_2H]. Significant results [11] have been obtained, however, using benzils with one labelled carbonyl carbon atom (see Scheme 6).
Here again it appears that electron-attracting substituents facilitate rearrangement, as evidenced by the following results.

Ar substituent	*p*-Me	*p*-OMe	*p*-Cl	*m*-Cl
% migration of Ar	38·8	31·8	67·2	81·2

The order of migratory aptitudes shown by these figures is the reverse of that assigned to groups in the pinacol rearrangement (p. 24). On the Ingold mechanism, the direction as well as the rate of rearrangement depends both on the reversible addition of HO^- and on the actual migration step. The above results become intelligible on the assumption that the predominant effect of a substituent is that exerted upon the first stage in the reaction (i.e., nucleophilic addition to the carbonyl group).

Rearrangement of tertiary α-ketols. This related transformation is responsible for anomalies observed in the synthesis of ketols by the reaction of benzils with Grignard reagents, and also in the alkaline degradation of the ketols. The following basic reactions are observed:

$$\mathrm{PhCOCOPh + PhMgBr \longrightarrow Ph_2C(OH)COPh}$$
$$\mathrm{Ph_2C(OH)COPh + KOH \longrightarrow Ph_2CHOH + KOCOPh}$$

Note that, in the latter reaction, $Ph_2\bar{C}(OH)$ is a fairly good anionic leaving-group; cf. benzhydryl phenyl ketone (Ph_2CHCOPh) with boiling ethanolic potassium hydroxide gives diphenylmethane and potassium benzoate. For the Grignard reaction, there is formed the intermediate complex (20), whose anion can undergo a reversible 'semi-benzilic' change as shown.

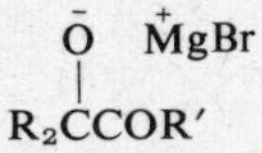

$$\underset{(20)}{\mathrm{R_2C(\bar{O}\overset{+}{M}gBr)COR'}} \qquad \mathrm{R_2C(O^-)\!-\!C(=O)R'} \rightleftharpoons \mathrm{RC(=O)\!-\!C(O^-)R'R}$$

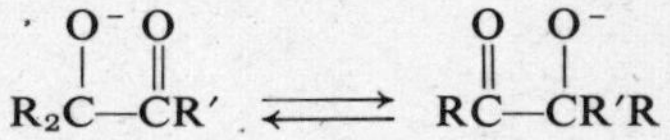

Consequently, starting with a given Grignard reagent (RMgBr) and diketone (RCOCOR′), a mixture of ketols may be formed. Since the same anion is inevitably produced, quickly and reversibly, when an α-ketol is treated with alkali, fission of an unsymmetrical ketol can yield two benzhydrols and two acids [12].

In contrast with the benzilic rearrangement, which is uni-directional owing to resonance stabilisation of the carboxyl function in the product, the ketol rearrangement is often reversible. It is not limited to aromatic compounds; it has been applied [13], in particular, to the enlargement of ring D in steroids (21).

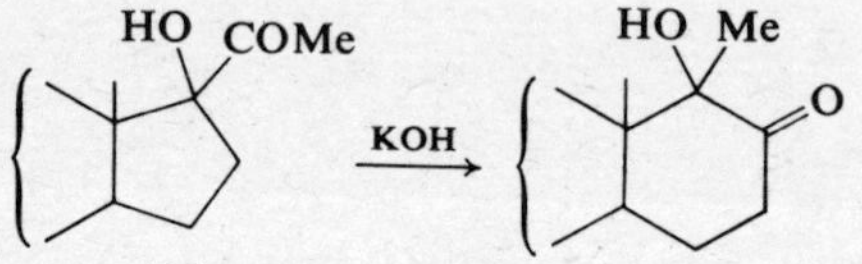

HO COMe — KOH → HO Me, O

(21)

Rearrangement of α-halo-ketones. These rearrangements are at least formally analogous to those of α-ketols, and a similar mechanism has been assigned (Scheme 7).

Scheme 7

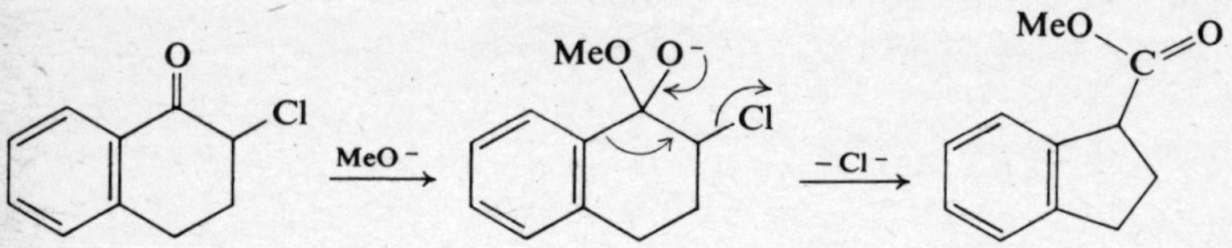

O, Cl — MeO^- → MeO O^-, Cl — $-Cl^-$ → MeO, C=O

Rearrangement of l-halo-l-aroylcyclohexanes gives l-arylcyclohexanecarboxylic acids, in one established case (22) with total stereochemical inversion at C_1. This finding agrees well with the proposed concerted mechanism but there is some doubt whether such a mechanism applies to all cases [14]. The transformation is often designated a 'quasi-Favorskiĭ' rearrangement (next section).

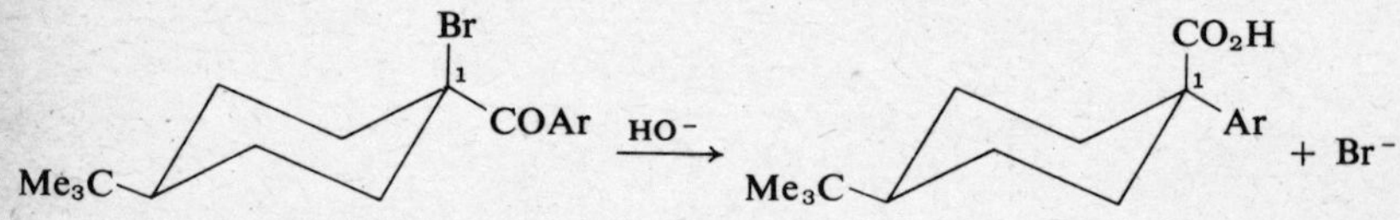

Me_3C, Br, 1, COAr — HO^- → Me_3C, CO_2H, 1, Ar + Br^-

(22)

Favorskiĭ and Related Rearrangements

The peculiar character of the three-membered ring (easily produced as well as easily opened) forms the basis of several molecular rearrangements of synthetic value. These can be based on the simple ring-closure:

$$\text{H—A}\overset{\text{B}}{\wedge}\text{C—X} \xrightarrow{\text{base}} \text{A}\overset{\text{B}}{\triangle}\text{C} \qquad (\text{X} = \text{halide or tosylate})$$

Such cyclisation is followed, often in one operation, by ring-opening in another sense.

Favorskiĭ rearrangement. The reaction of α-halo-ketones (23) with base, especially hydroxide or alkoxide, can take various courses:

(a) apparently simple displacement of halide by alkoxide (24);

(b) *C*-alkylation of one molecule by another, giving (25);

(c) formation of an alkoxy-epoxide (26) or a derived hydroxy-ketal (27);

(d) formation of an acid (28) with skeletal rearrangement.

This last transformation is the 'semi-benzilic' or 'quasi-Favorskiĭ' rearrangement just discussed.

RCH(OAlk)COR′ (24) ⟵ RCH(Cl)COR′ (23) ⟶ RCH(COR′)–RC(Cl)COR′ (25)

(23) ⟶ RCH(Cl)–C(R′)(OAlk)–O⁻ ⟶ RCH—C(R′)—OAlk epoxide (ring O) (26) ⟶ RCH(HO)—C(R′)(OAlk)—OAlk (27)

(23) ⟶ RR′CHCO₂H (28)

From 1894 onwards Favorskiĭ studied [15] a further complication, which can arise when the chlorine-free carbon atom adjoining the carbonyl group carries a hydrogen atom.

$$MeCOCMe_2Cl \longrightarrow HOCOCMe_3$$

$$\underset{(29)}{PhCH_2COCH_2Cl} \longrightarrow PhCH_2CH_2CO_2H \longleftarrow \underset{(30)}{PhCHClCOMe}$$

Many of these rearrangements can be accommodated by the semi-benzilic mechanism discussed previously in connection with 'quasi-Favorskiĭ' changes. Richard proposed the following scheme involving an intermediate carbene and ketene.

$$\begin{array}{c}R-C{=}O\\|\\R-CHCl\end{array} \xrightarrow{-HCl} \left[\begin{array}{c}R-C-O^-\\\|\\R-C^+\end{array} \longleftrightarrow \begin{array}{c}R-C{=}O\\|\\R-C:\end{array}\right] \longrightarrow \begin{array}{c}O\\\|\\C\\\|\\R_2C\end{array} \xrightarrow{H_2O} \begin{array}{c}O\\\|\\C-OH\\|\\R_2CH\end{array}$$

Favorskiĭ himself regarded the alkoxy-epoxide (26) as an intermediate. However, in several cases, these have been prepared independently and found to be stable under the rearrangement conditions. None of these suggested mechanisms will account for the formation of the same acid from (29) and (30) (*vide supra*).

After an early suggestion of Wallach, a cyclopropanone intermediate was postulated.

$$\begin{array}{l}PhCH_2COCH_2Cl \searrow\\PhCHClCOCH_3 \nearrow\end{array} \begin{array}{c}O\\C\\/\ \backslash\\PhCH-CH_2\end{array} \longrightarrow PhCH_2CH_2CO_2H$$

In this intermediate, the two carbon atoms α- to the carbonyl group are similarly situated, so that the above transformations of (29) and (30) are accommodated. The few cyclopropanones which are known are unstable, and very easily undergo the postulated ring-opening. This involves at least potentially the detachment of one α-carbon atom from the carbonyl group as an anion, which will be favoured if this atom is arylated, as in the rearrangement of (29) and (30), and disfavoured by alkylation (cf. reaction of $MeCOCMe_2Cl$ above).

The need to invoke a cyclopropanone or similar quasi-symmetrical intermediate was neatly emphasised [16] by Loftfield. Rearrangement of 2-chlorocyclohexanone (31) (see Scheme 8), labelled nearly equally at C_1 and C_2, yielded cyclopentanecarboxylic acid of full activity.

Scheme 8

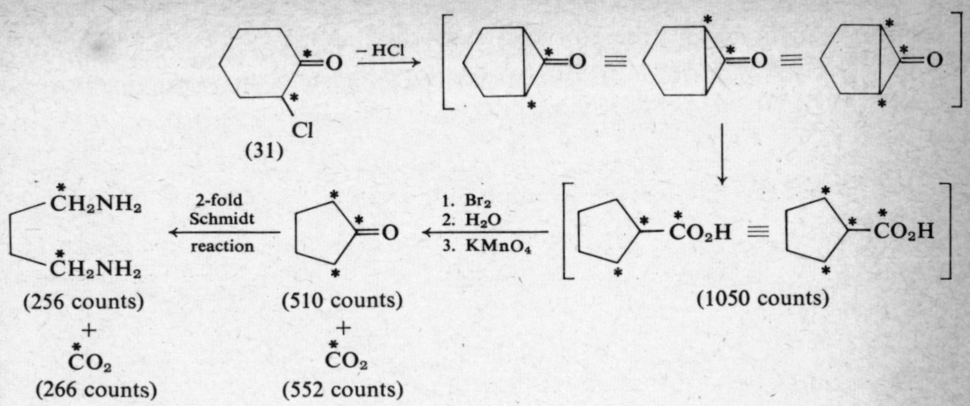

(31)

(256 counts) + $\overset{*}{C}O_2$ (266 counts)

(510 counts) + $\overset{*}{C}O_2$ (552 counts)

(1050 counts)

From earlier prevailing views of the mechanism, it would be expected that this activity would be divided equally between the carboxyl carbon atom and the attached ring carbon atom. Degradation of the molecule disclosed a different picture, consonant with the cyclopropanone mechanism. Note that, in the cyclopropanone, the bridgehead carbon atoms are identically situated and each must be shown as labelled, carrying half the activity of the carboxyl carbon atom; similar considerations apply to the final cyclopentanecarboxylic acid. It appears that, when the cyclopropanone mechanism or its equivalent is available (presence of α-halogen and α′-halogen), the reaction will normally follow this course rather than the semi-benzilic or Richard sequence, unless the initial material has a cyclic structure unfavourable to formation of a cyclopropanone ring [17].

Dewar has pointed out [18] that the ring-closure

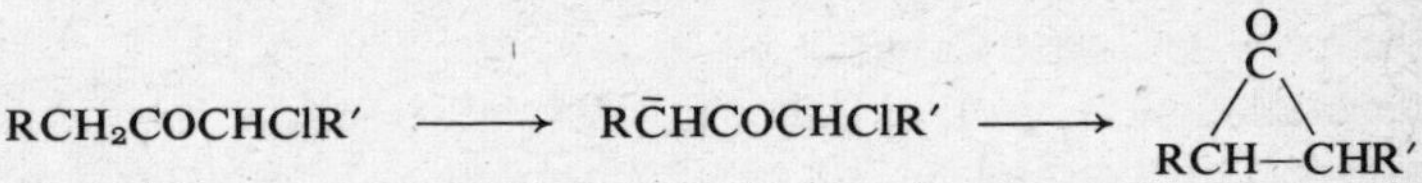

$$RCH_2COCHClR' \longrightarrow R\bar{C}HCOCHClR' \longrightarrow \text{RCH—CHR}' \text{ (cyclopropanone, C=O)}$$

is awkward stereoelectronically and, following Aston, has suggested that the intermediate is not primarily the complete cyclopropanone but a symmetrical, highly mesomeric entity, partially represented as

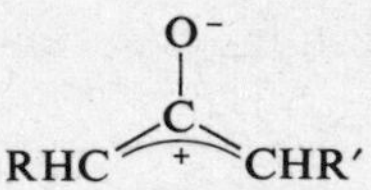

$$\text{RHC} \overset{\overset{O^-}{|}}{\underset{+}{\overset{C}{\frown}}} \text{CHR}'$$

(This structure, incidentally, could be interconverted with Richard's proposed carbene intermediate by transfer of a hydride ion.) The complete sequence could therefore be as follows:

α-halo-ketone $\longrightarrow$ enolate $\longrightarrow$ Aston–Dewar intermediate $\longrightarrow$ cyclopropanone $\longrightarrow$ Favorskiĭ product.

In several instances, it has been shown by deuterium-exchange experiments that reversible deprotonation of the ketone precedes rearrangement.

Scheme 9

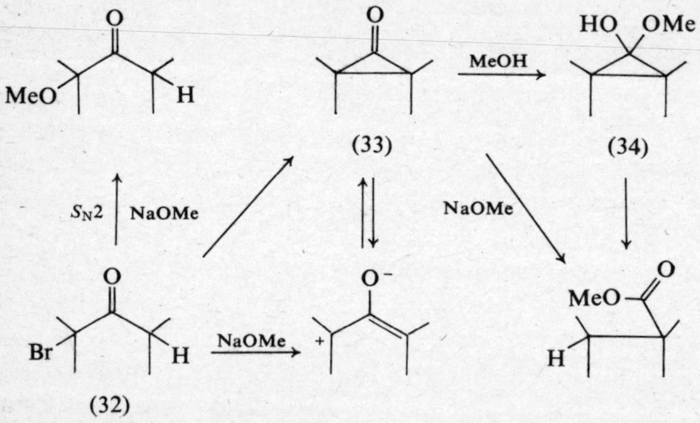

Turro suggested [19] the probable scheme of reactions (Scheme 9), having studied the behaviour of α-bromo-diisopropyl ketone (32) and also the related tetramethylcyclopropanone (33) and its methyl semi-acetal (34). He also showed that four ketones, namely the α- and α'-bromo- and -chloro-derivatives of ethyl isopropyl ketone, each gave the same mixture of products as did trimethylcyclopropanone under the same conditions—notable evidence for the intermediacy of the cyclopropanone in this series.

In a detailed study [20] of the particular substrate $ArCH_2COCH_2Cl$, Bordwell observed that the reaction is accelerated by electron-donating substituents in Ar (Hammett ρ-value $= -5$) and concluded that the production of the Aston–Dewar intermediate from the enolate was the probable rate-determining step; the kinetic study indicated that removal of the proton and of chloride were independent. The large negative ρ-value tells against a direct conversion of enolate into cyclopropanone, since the analogous reaction of α-halo-sulphones (next section), where direct three-membered ring-formation is the accepted mechanism, gives the value $\rho = -2{\cdot}8$. Parallel experiments with the homologous ketone $ArCH_2COCHMeCl$ gave $\rho = +0{\cdot}95$ and otherwise indicated that, in this case, the second stage is so rapid that proton-removal is significant in rate-determination.

It was also observed that formation of the methoxy-ketone $ArCH_2COCHMeOMe$ is the main reaction of the last-mentioned ketone at low temperature and low methoxide concentration. Bordwell concludes that the *free* enol reacts with methanol (cf. cinnamyl halides undergo nucleophilic substitution with outstanding ease).

An important stereochemical point arises. The direct internal S_N2 conversion of enolate into cyclopropanone should be stereospecific with inversion about the halogen-bearing carbon atom, whereas reaction through the Aston–Dewar intermediate is stereochemically undefined. In practice, the rearrangement of (35) gives (36) stereospecifically when carried out in ether or dimethoxyethane, but is non-stereospecific in methanol. It is suggested that the more-ionising solvent favours the production of the Aston–Dewar intermediate. Production of the cyclopropanone directly and via the Aston–Dewar intermediate are, in essence, respectively cases of S_N2 and S_N1 displacement.

COMe, Cl, H, Me → Me, CO_2H, H, Me

(35) (36)

Experiments with steroid halo-ketones, however, led Engel to the conclusion [21] that not only the incidence of the Aston–Dewar mechanism but also its stereochemical consequences may be determined both by the reaction conditions and by the steric environment of the reactive system in a relatively rigid polycyclic molecule.

The Favorskiĭ rearrangement is also observed in α,α- and α,α'-dihalo-ketones, again apparently proceeding through a three-membered ring intermediate [22] and yielding an unsaturated acid.

Synthetic uses. The rearrangement can be effected reasonably smoothly with a wide range of halo-ketones, the chloro-compounds proving best. The reaction has been much used with alicyclic compounds, especially steroids. Thus, 2-α-bromocholestan-3-one suffers contraction of ring A [23] and the method can be used to insert a methyl group in ring D at C_{17} [21, 24].

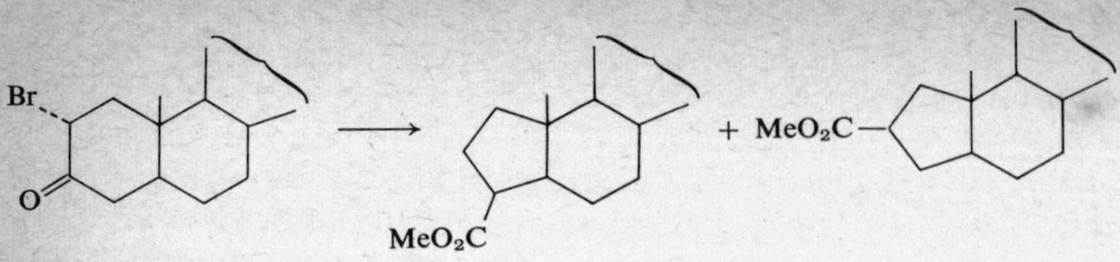

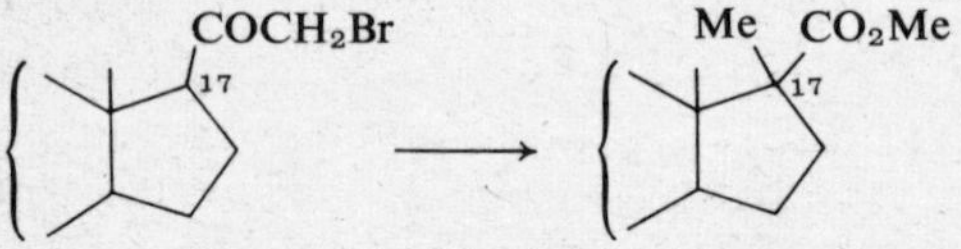

Rearrangement of halo-sulphones. Ramberg and Bäcklund showed [25] that an α-halo-sulphone ($RCHXSO_2CH_2R'$; X = halogen), when warmed with alkali, can yield the alkene $RCH{=}CHR'$, along with halide and sulphite ions. The reaction probably proceeds by way of an episulphone ($\overset{SO_2}{\overbrace{RCH{-}CHR'}}$), corresponding to the cyclopropanone intermediate of the Favorskiĭ rearrangement. Formation of a carbene ($R\ddot{C}SO_2CH_2R'$) is not involved since tertiary halo-sulphones react in the same way as the secondary type.

$$\underset{(37)}{MeCHBrSO_2CH_2Me} \xrightarrow{-HBr} \underset{(38)}{\underset{\underset{O_2}{S}}{MeCH{-}CHMe}} \xrightarrow{-SO_2} \underset{(39)}{MeC_{(2)}H{=}C_{(3)}HMe}$$

Bromo-diethyl sulphone (37) has been studied in detail [26]. The postulated intermediate episulphone (38) has been prepared from diazoethane and sulphur dioxide, and the pure *cis*-form separated. With aqueous potassium hydroxide (2N), the sulphone (37) gives the alkene (39), more than 75 per cent *cis*, while the episulphone (38) gives purely *cis*-(39). With KOD/D_2O, (37) gives (39) with 97 per cent deuteration at C_2 and C_3 although, under similar conditions, (38) gives a product only 5 per cent deuterated. With the much more strongly basic potassium t-butoxide in t-butanol, the product from (37) or (38) is 75 per cent *trans*-(39), fully deuterated at C_2 and C_3 in each case.

The reaction sequence appears to involve rapid reversible deprotonation of (37) followed by formation of (38), mainly the *cis*-isomer; when the strong base is used, epimerisation of (38) by reversible deprotonation gives an equilibrium mixture favouring the *trans*-form, and thence formation of mainly *trans*-(39). In the less alkaline aqueous medium, the now much slower stereochemical equilibration of (38) is outstripped by the rate of loss of SO_2.

Kinetic studies [27] of this rearrangement indicate a rapid reversible deprotonation step (since sulphones are very weakly acidic, the equilibrium lies far on the side of the neutral species, i.e., $>CHSO_2R \rightleftarrows >\bar{C}SO_2R$), followed by rate-determining ring-closure and loss of SO_2. The rate of rearrangement increases through the series: $MeSO_2CH_2Cl$ < $MeSO_2CHPhCl \ll PhCH_2SO_2CH_2Cl < PhCH_2SO_2CHPhCl$. The mechanism of the elimination of SO_2 (i.e., 38 → 39) is still *sub judice*.

Scheme 10

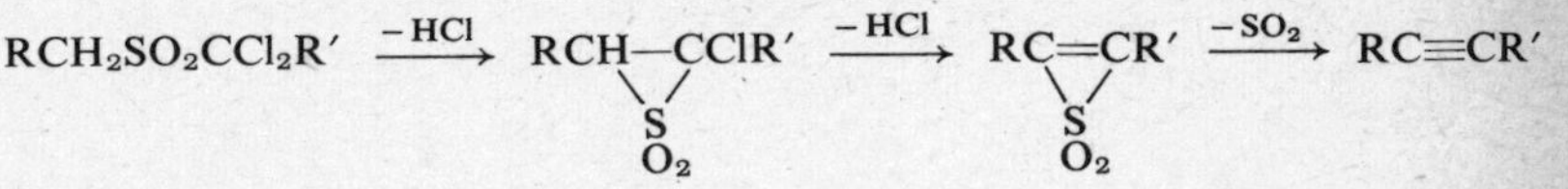

$$RCH_2SO_2CCl_2R' \xrightarrow{-HCl} \underset{(40)}{RCH{-}CClR'\ (\text{ring } SO_2)} \xrightarrow{-HCl} \underset{(41)}{RC{=}CR'\ (\text{ring } SO_2)} \xrightarrow{-SO_2} RC{\equiv}CR'$$

The behaviour of dichloro-sulphones has been formulated as shown in Scheme 10. The sulphonate anions, $RCH{=}CR'SO_3^-$ and $R'CH{=}CRSO_3^-$ are also formed, derivable [28] from (40) and (41). Triethylamine reacts with α α′-dibromodibenzyl sulphone to give the thiirene dioxide which can yield both diphenylacetylene and stilbenesulphonate anion.

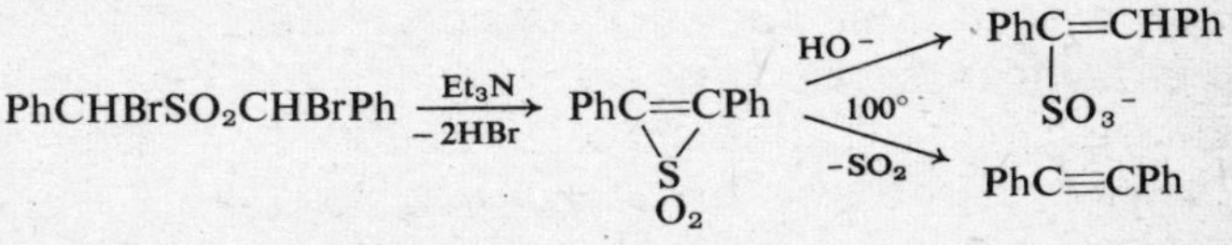

$$PhCHBrSO_2CHBrPh \xrightarrow[-2HBr]{Et_3N} PhC{=}CPh\ (\text{ring } SO_2) \begin{cases} \xrightarrow{HO^-} PhC(SO_3^-){=}CHPh \\ \xrightarrow[-SO_2]{100^\circ} PhC{\equiv}CPh \end{cases}$$

Synthetic uses. The reaction offers a practical means of homologation of alkenes:

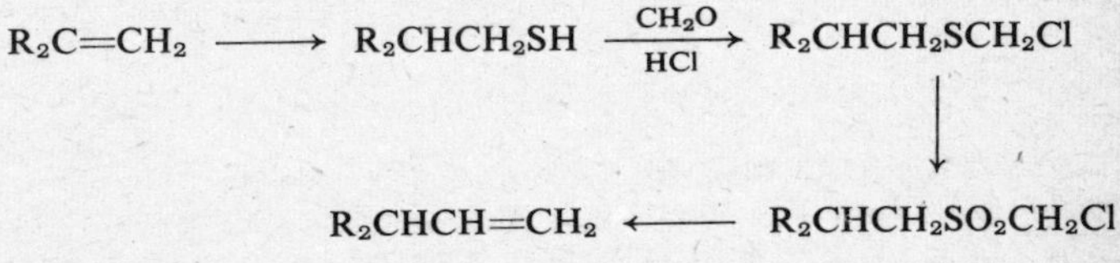

$$R_2C{=}CH_2 \longrightarrow R_2CHCH_2SH \xrightarrow[HCl]{CH_2O} R_2CHCH_2SCH_2Cl$$

$$\downarrow$$

$$R_2CHCH{=}CH_2 \longleftarrow R_2CHCH_2SO_2CH_2Cl$$

A more elaborate sequence [29] can give a long-chain unsaturated carboxylic acid (see Scheme 11).

Scheme 11

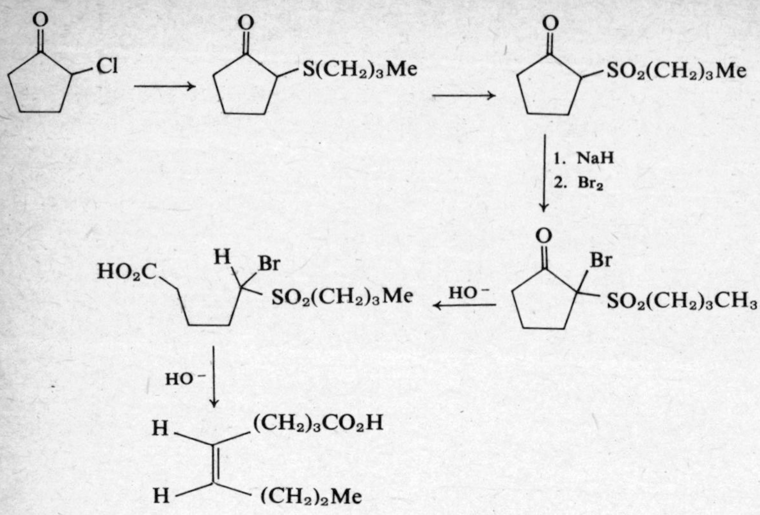

2-chlorocyclopentanone $\longrightarrow$ 2-[$S(CH_2)_3Me$]cyclopentanone $\longrightarrow$ 2-[$SO_2(CH_2)_3Me$]cyclopentanone

$\downarrow$ 1. NaH 2. Br_2

2-bromo-2-[$SO_2(CH_2)_3CH_3$]cyclopentanone $\xrightarrow{HO^-}$ $HO_2C(CH_2)_3CHBr\,SO_2(CH_2)_3Me$

$\downarrow HO^-$

$H(CH_2)_3CO_2H\,C{=}C\,H(CH_2)_2Me$

The alkaline ring-opening of the cyclopentanone derivative in this reaction sequence is an example of the very easy 'acid hydrolysis' of β-oxo-sulphones.

Neber reaction. In a series of papers (1926–36), Neber reported [30] an unexpected reaction of the toluenesulphonyl esters of ketoximes which, in its final effect, involves migration of nitrogen. Dependent on the reaction conditions and method of working up, the ester with potassium ethoxide can give an amino-ketal, the amino-ketone hydrochloride, or a pyrazine (the known product of ready cyclisation and dehydrogenation of α-amino-ketones).

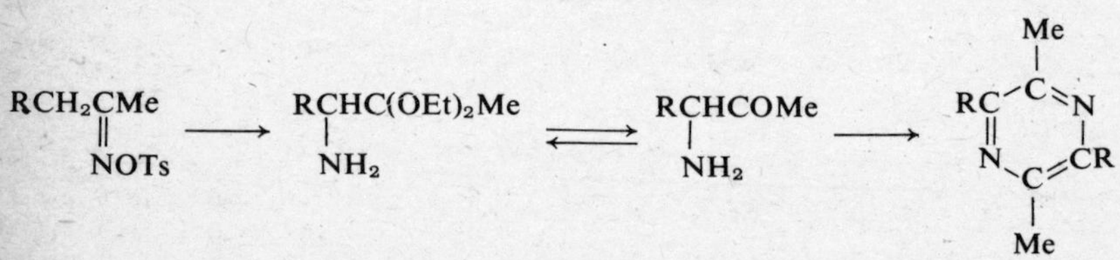

$$RCH_2C(=NOTs)Me \longrightarrow RCH(NH_2)C(OEt)_2Me \rightleftharpoons RCH(NH_2)COMe \longrightarrow \text{2,5-dimethyl-3,6-R}_2\text{-pyrazine}$$

The reaction often affords good yields of amino-ketones, and has been very serviceable in cases where the usual method of inserting an α-amino-group into a ketone by way of the oximino-compound has been unsatisfactory. It constitutes a key step in the synthesis of lysergic acid [31]. Occasionally, however, the main reaction is Beckmann rearrangement of the oxime ester. Aldoxime tosylates give only the nitrile.

Scheme 12

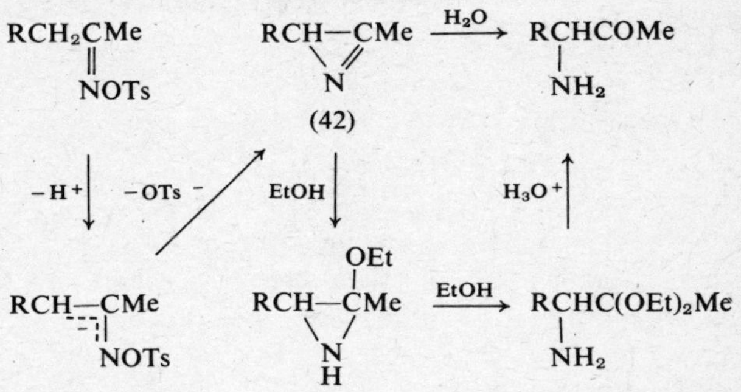

$$RCH_2C(=NOTs)Me \xrightarrow{-H^+} RCH{-}C(\text{-}{-}NOTs)Me \xrightarrow{-OTs^-} \underset{(42)}{RCH{-}CMe\ (\text{azirine}, N)} \xrightarrow{H_2O} RCH(NH_2)COMe$$

$$(42) \xrightarrow{EtOH} RCH{-}C(OEt)Me\ (\text{aziridine}, NH) \xrightarrow{EtOH} RCH(NH_2)C(OEt)_2Me \xrightarrow{H_3O^+} RCH(NH_2)COMe$$

A three-membered ring (azirine) intermediate (42) appears to be involved in the reaction (see Scheme 12). This mechanism was advocated [32] by Cram on the basis of experiments with benzyl phenyl ketoxime tosylate, in which the structure of the isolated ethoxyaziridine was confirmed by its spectrum and by its reduction by lithium aluminium hydride to the known *cis*-2,3-diphenylaziridine.

In related experiments [33] with sec. alkyldichloroamines, the intermediate azirine was similarly recognised:

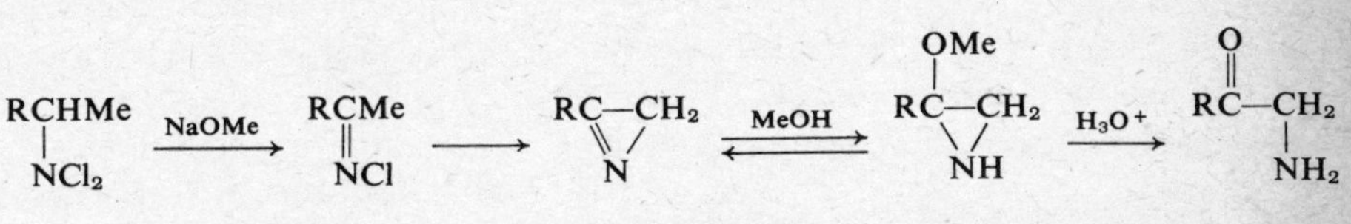

$$RCH(NCl_2)Me \xrightarrow{NaOMe} RC(=NCl)Me \longrightarrow RC{-}CH_2\ (\text{azirine}, N) \underset{}{\overset{MeOH}{\rightleftharpoons}} RC(OMe){-}CH_2\ (\text{aziridine}, NH) \xrightarrow{H_3O^+} RC(=O){-}CH_2NH_2$$

As well as TsO^- and Cl^-, Me_3N is a possible leaving-group (Parcell):

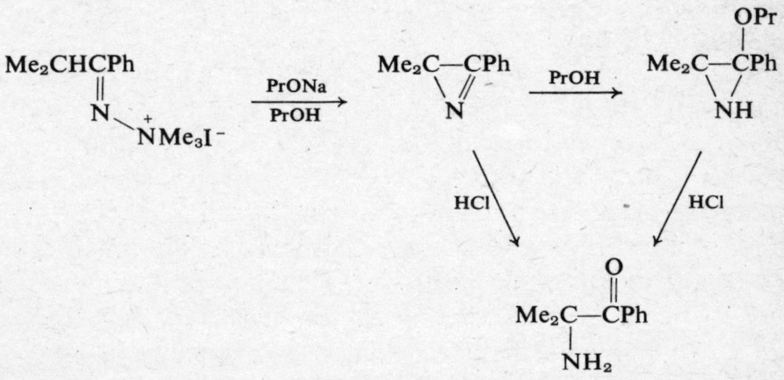

$$Me_2CHC(Ph)=N{-}\overset{+}{N}Me_3I^- \xrightarrow[PrOH]{PrONa} Me_2C{-}CPh\ (\text{azirine}, N) \xrightarrow{PrOH} Me_2C{-}C(OPr)Ph\ (\text{aziridine}, NH)$$

$$\text{both} \xrightarrow{HCl} Me_2C(NH_2){-}C(=O)Ph$$

The Chapman and Related Rearrangements

These rearrangements are of the type:

$$\underset{\substack{|\\R}}{X}-Y=Z \longrightarrow \text{X}\overset{Y}{\cdots}\text{Z (cyclic, via R)} \longrightarrow X=Y-\underset{\substack{|\\R}}{Z}$$

Many reactions of this sort are regarded as proceeding via a four-centre transition state as indicated. Y is usually carbon and X and Z more electronegative elements (N, O, S), so that the compounds concerned are related to carboxylic acids. The changes are formally allied, rather distantly, to the typical electrophilic migrations, in that R proceeds to a lone pair of electrons on Z.

Imino-ether rearrangements. Chapman observed [34] that aryl imino-ethers, readily obtainable from imino-chlorides and alkali phenoxides, underwent rearrangement at about 220–250°:

$$PhC(=O)NHPh \xrightarrow{PCl_5} PhC(Cl)=NPh \xrightarrow{PhO^-} PhC(OPh)=NPh \xrightarrow{heat} PhC(=O)NPh_2$$

The reaction proceeds smoothly in the absence of thermo-labile substituents, and affords a practical route to diarylamines, especially the unsymmetrical variety.

The method has been improved [35] by using thionyl chloride in the first stage and by conducting the final isomerisation in boiling tetramethylene glycol dimethyl ether. It has been applied successfully to the synthesis of diphenylamines that have nuclear halogeno-, methoxy-, cyano-, aryl-, or ethoxycarbonyl-substituents, and, in one or two cases, even a nitro-substituent.

An important special case is the use [36] of a salicylic ester as phenolic starting material, leading to a versatile synthesis of acridones as shown in Scheme 13.

Kinetically [33, 37] the reaction is first-order. The rather small effect of nuclear substitution is in accord with the mechanistic formulation as a four-centre cyclic process. Consistent with the overall result (i.e., migration of aryl to a nitrogen lone pair), however, the process is promoted by electron-attracting substituents in the migrating aryl group, and retarded to a lesser extent by such substituents in the stationary aryl group attached to nitrogen (and also in the *C*-aryl group). Mixture experiments indicate that the process is intramolecular [34, 35, 37].

Scheme 13

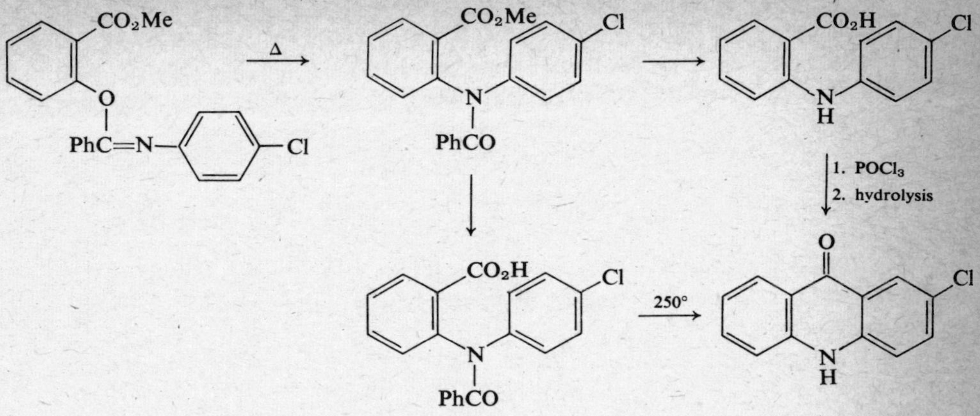

An equally smooth migration of alkyl groups under the same conditions is not general, although one long-known case is the conversion of cyanuric esters into isocyanurates. Migration of an aryl group allows the representation (43) for the four-membered ring intermediate.

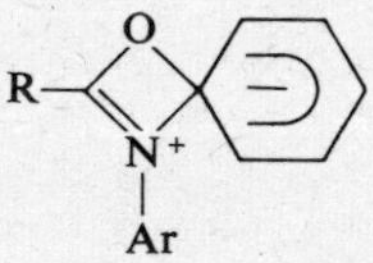

(43)

Allyl groups, on the other hand, can migrate freely with 'inversion' [35, 38] (e.g., $—OCH_2CH{=}CHEt \rightarrow —NCHEtCH{=}CH_2$). The analogous isomerisation of crotyl 2-quinolyl ether (44) emphasises the parallel with the Claisen rearrangement and other sigmatropic processes (see Chapter 8).

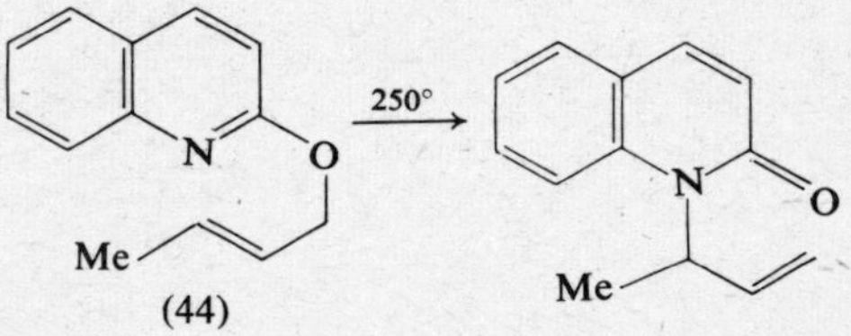

(44)

Rearrangement of alkyl imino-ethers, however, can be effected by heating with small quantities of alkyl halide or with an acid reagent, such as boron trifluoride or sulphuric acid, capable of yielding an alkyl cation. The changes appear to be intermolecular and may be formulated as follows:

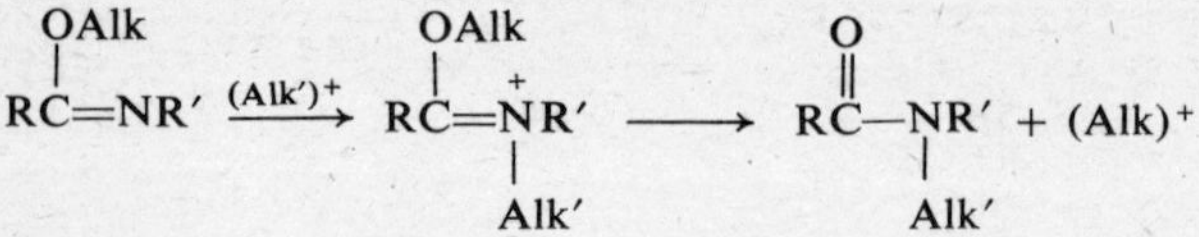

$$\mathrm{RC(OAlk){=}NR'} \xrightarrow{(\mathrm{Alk'})^+} \mathrm{RC(OAlk){=}\overset{+}{N}R'(Alk')} \longrightarrow \mathrm{RC({=}O){-}NR'(Alk')} + (\mathrm{Alk})^+$$

(Compare the Arbuzov reaction, p. 116). A sequence (Scheme 14) has been developed [39] on this basis for the mono-alkylation of primary aromatic amines.

Scheme 14

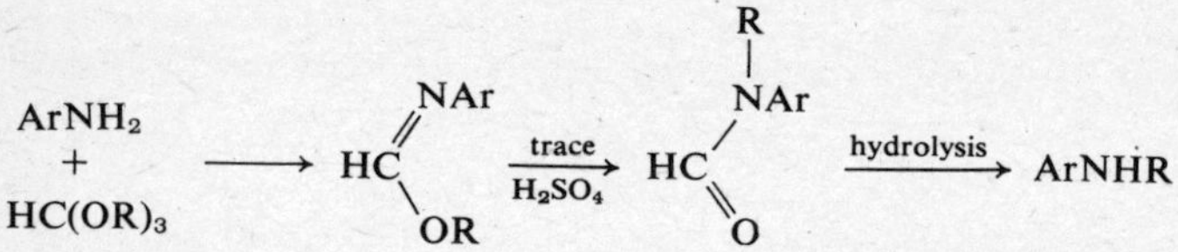

$$\mathrm{ArNH_2} + \mathrm{HC(OR)_3} \longrightarrow \mathrm{HC({=}NAr)OR} \xrightarrow[\mathrm{H_2SO_4}]{\text{trace}} \mathrm{HC({=}O)N(R)Ar} \xrightarrow{\text{hydrolysis}} \mathrm{ArNHR}$$

Acyl groups can show a similar migration [40] very easily, as shown by (45).

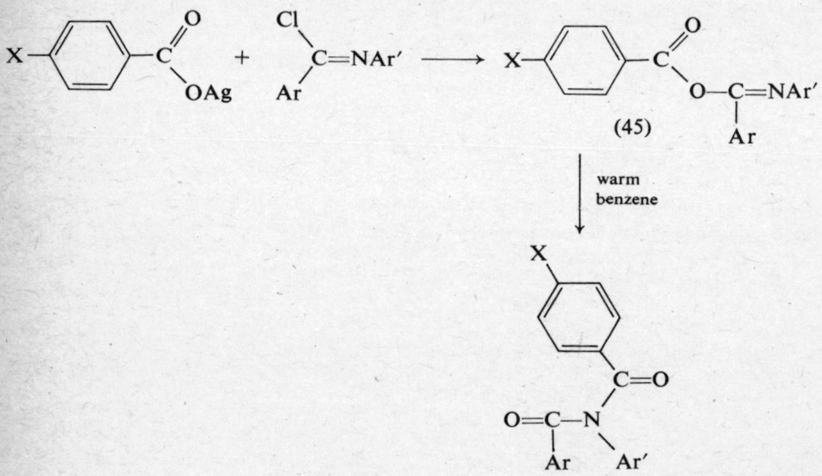

$$\mathrm{X{-}C_6H_4{-}C({=}O)OAg} + \mathrm{Ar{-}C(Cl){=}NAr'} \longrightarrow \underset{(45)}{\mathrm{X{-}C_6H_4{-}C({=}O){-}O{-}C(Ar){=}NAr'}} \xrightarrow{\text{warm benzene}} \mathrm{X{-}C_6H_4{-}C({=}O){-}N(Ar'){-}C({=}O){-}Ar}$$

When Ar′ = Ph, the intermediate ‘imino-anhydride’ cannot be isolated. For Ar′ = 2,4-$(O_2N)_2C_6H_3$ this can be achieved, and the second (rearrangement) stage shows first-order kinetics and moderate acceleration when X is electron-attracting. The four-centre mechanism is advocated and requires that, in the imino-anhydride, Ar′ be *syn* to OAcyl; in phenylimino-phthalide (46), this is sterically impossible, and rearrangement to phthalanil is slow even at 250°.

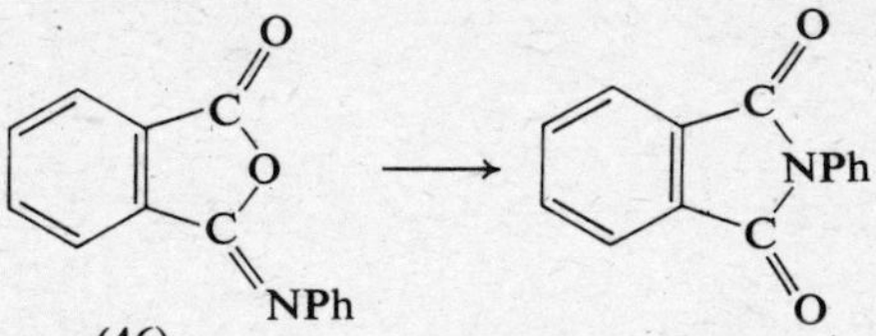

(46)

Chapman has also recorded [34] a rearrangement of imino-thioethers, more difficult to effect than with the imino-ethers, and apparently reversible.

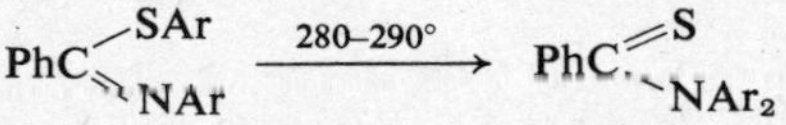

$$\mathrm{PhC}\begin{matrix}\nearrow \mathrm{SAr} \\ \searrow \mathrm{NAr}\end{matrix} \xrightarrow{280\text{–}290^\circ} \mathrm{PhC}\begin{matrix}\nearrow \mathrm{S} \\ \searrow \mathrm{NAr_2}\end{matrix}$$

Amidine rearrangement. Triarylamidines rearrange in the same way as imino-ethers (Chapman [41]).

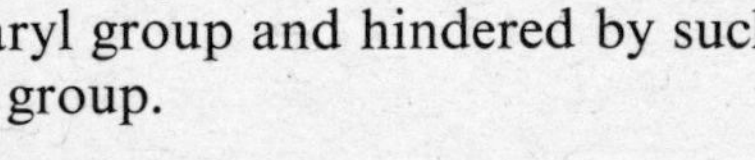

$$\mathrm{PhC}\begin{matrix}\nearrow \mathrm{NArAr'} \\ \searrow \mathrm{NAr'}\end{matrix} \xrightarrow{300^\circ} \mathrm{PhC}\begin{matrix}\nearrow \mathrm{NAr} \\ \searrow \mathrm{NAr'_2}\end{matrix}$$

A higher temperature is required than for the previous cases and the reaction is reversible. Like the imino-ether rearrangement, it is an intramolecular process, promoted by electron-attracting substituents in the migrating aryl group and hindered by such substituents in the stationary *N*-aryl group.

Oxygen to sulphur migrations. Rearrangements of the type shown below are observed [42] in molecules related to carbonic acid with X = OR, SR, or NR_2.

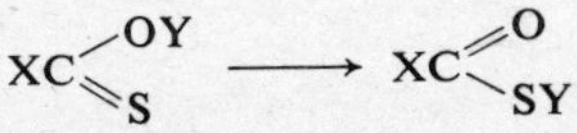

$$\mathrm{XC}\begin{matrix}\nearrow \mathrm{OY} \\ \searrow \mathrm{S}\end{matrix} \longrightarrow \mathrm{XC}\begin{matrix}\nearrow \mathrm{O} \\ \searrow \mathrm{SY}\end{matrix}$$

In a series of migrations where Y = alkyl, reaction was found to be promoted by electron-attracting power in X. When Y = aryl, the rearrangement requires a high temperature, as in the Chapman rearrangement, but allyl-type groups migrate slowly at room temperature with 'inversion'.

Rearrangements of cyanates, thiocyanates, and nitrile oxides. The 1,3-migrations from oxygen or sulphur to nitrogen in the first two classes are perhaps only formally associated with the changes just discussed. Alkyl cyanates, effectively isolated quite recently, rearrange on warming to give isocyanates, with some trimer:

$$RO{-}C{\equiv}N \xrightarrow{100^\circ} O{=}C{=}NR$$

Attempted preparation of allyl cyanates gave the isocyanate directly.

Alkyl thiocyanates are much more stable, giving isothiocyanates at about 170°, while the allyl compound rearranges to 'mustard oil' on gentle warming. Crotyl and similar thiocyanates rearrange with inversion:

$$\begin{array}{c} MeCH{=}CHCH_2 \\ | \\ N{\equiv}C{-}S \end{array} \longrightarrow \begin{array}{c} MeCHCH{=}CH_2 \\ | \\ N{=}C{=}S \end{array}$$

Speed of reaction is not sensitive to change in polarity of medium, and the process is regarded as a six-centre sigmatropic change. A detailed study [43] of diphenylmethyl thiocyanate, Ph_2CHSCN, showed that reaction is favoured strongly by polar solvents, by electron-donating substituents in the benzene rings, and by addition of salts. Together with kinetic measurements, including investigation of the effect of added labelled thiocyanate ($*SCN^-$), these results lead to the conclusion that the main process of rearrangement of saturated alkyl thiocyanates via an intimate ion-pair ($R^+ \; SCN^-$) is here superseded by a displacement:

$$*SCN^- + RSCN \longrightarrow *SCNR + SCN^-$$

A very curious rearrangement of a cyanogen derivative, perhaps not really 'belonging here', is the thermal transformation of nitrile oxides (which are *C*-alkyl fulminates!) into isocyanates [44]:

$$R{-}C{\equiv}\overset{+}{N}{-}\overset{-}{O} \longrightarrow O{=}C{=}N{-}R$$

The reaction is intramolecular and does not involve the formation of R^+, since no sign of Wagner–Meerwein change is detected when, for example, R = camphenilyl. It is suggested that isomerisation proceeds through a transition state formulable as shown below.

References

1. K. C. Brannock, H. Pridgen, and B. Thompson, *J. Org. Chem.*, 1960, **25**, 1815.
2. U. Schöllkopf and I. Hoppe, *Tetrahedron Letters*, 1970, 4527.
3. S. Smiles *et al.*, *J. Chem. Soc.*, 1931, 2207; 1935, 181, 1234 etc.
4. J. F. Bunnett and T. Okamoto, *J. Amer. Chem. Soc.*, 1956, **78**, 5363.
5. N. W. Gilman, P. Levitan, and L. H. Sternbach, *Tetrahedron Letters*, 1970, 4121.
6. W. E. Truce *et al.*, *J. Amer. Chem. Soc.*, 1958, **80**, 3625.
7. W. E. Truce and W. W. Brand, *J. Org. Chem.*, 1970, **35**, 1828.
8. I. Roberts and H. C. Urey, *J. Amer. Chem. Soc.*, 1938, **60**, 880.
9. W. von E. Doering and R. S. Urban, *J. Amer. Chem. Soc.*, 1956, **78**, 5938.
10. C. K. Ingold, *Ann. Reports*, 1928, **25**, 124.
11. M. T. Clark, E. C. Hendley, and O. K. Neville, *J. Amer. Chem. Soc.*, 1955, **77**, 3280.
12. J. F. Eastham *et al.*, *J. Amer. Chem. Soc.*, 1956, **78**, 4323.
13. C. W. Shoppee and D. A. Prins, *Helv. Chim. Acta*, 1943, **26**, 186.
14. B. Tchoubar, *Bull. Soc. chim. France*, 1955, 1363; D. Baudry, J. F. Bégué and M. Charpentier-Morize, *ibid.*, 1971, 1416.
15. A. E. Favorskiĭ, *J. Russ. Phys.-Chem. Soc.*, 1894, **26**, 559.
16. R. B. Loftfield, *J. Amer. Chem. Soc.*, 1951, **73**, 4707.
17. E. W. Warnhoff, C. M. Wong, and W. T. Tai, *J. Amer. Chem. Soc.*, 1968, **90**, 514; J.-M. Conia and J.-L. Ripoli, *Bull. Soc. chim. France*, 1963, 773.
18. J. G. Burr and M. J. S. Dewar, *J. Chem. Soc.*, 1954, 1201.
19. N. J. Turro *et al.*, *J. Amer. Chem. Soc.*, 1965, **87**, 3258; 1970, **92**, 2032.
20. F. G. Bordwell *et al.*, *J. Amer. Chem. Soc.*, 1969, **91**, 2087, 3949.
21. C. R. Engel *et al.*, *Canad. J. Chem.*, 1970, **48**, 361.
22. R. Breslow *et al.*, *J. Amer. Chem. Soc.*, 1965, **87**, 1326.
23. C. W. Shoppee *et al.*, *J. Chem. Soc.*, 1957, 1451.
24. P. A Plattner, H. Heusser, and S. F. Boyce, *Helv. Chim. Acta*, 1948, **31**, 603.
25. L. Ramberg and B. Bäcklund, *Arkiv Kemi Min. Geol.*, 1940, **13A**, no. 27; see *Chem. Abstr.*, 1940, **34**, 4725.
26. N. P. Neureiter, *J. Amer. Chem. Soc.*, 1966, **88**, 558.

27. F. G. Bordwell *et al.*, *J. Amer. Chem. Soc.*, 1951, **73**, 5187; 1968, **90**, 429, 435.
28. L. A. Paquette, L. S. Wittenbrook, and V. V. Kane, *J. Amer. Chem. Soc.*, 1967, **89**, 4483, 4487; L. A. Carpino *et al.*, *ibid.*, 1971, **93**, 476.
29. J. Ficini and G. Stork, *Bull. Soc. chim. France*, 1964, 723.
30. P. W. Neber *et al.*, *Annalen*, 1926, **449**, 109 *et seq.*
31. R. B. Woodward *et al.*, *J. Amer. Chem. Soc.*, 1956, **78**, 3087.
32. D. J. Cram and M. J. Hatch, *J. Amer. Chem. Soc.*, 1953, **75**, 33, 38.
33. H. E. Baumgarten *et al.*, *J. Amer. Chem. Soc.*, 1954, **76**, 4561 *et seq.*
34. A. W. Chapman, *J. Chem. Soc.*, 1925, 1992; 1926, 2296; 1927, 1743.
35. O. H. Wheeler *et al.*, *Canad. J. Chem.*, 1969, **47**, 503; *J. Org. Chem.*, 1969, **34**, 966.
36. M. M. Jamison and E. E. Turner, *J. Chem. Soc.*, 1937, 1954.
37. K. B. Wiberg and B. I. Rowland, *J. Amer. Chem. Soc.*, 1955, **77**, 2205.
38. W. M. Lauer and C. S. Benton, *J. Org. Chem.*, 1959, **24**, 804.
39. R. M. Roberts and P. J. Vogt, *J. Amer. Chem. Soc.*, 1956, **78**, 4778.
40. D. Y. Curtin and L. L. Miller, *J. Amer. Chem. Soc.*, 1967, **89**, 637.
41. A. W. Chapman, *J. Chem. Soc.*, 1929, 2133 *et seq.*
42. C. Christophersen and A. Holm, *Acta Chem. Scand.*, 1970, **24**, 1512.
43. A. Fava, A. Iliceto, and S. Bresadola, *J. Amer. Chem. Soc.*, 1965, **87**, 4791.
44. C. Grundmann and P. Kochs, *Angew. Chem. Internat. Edn.*, 1970, **9**, 635.

REVIEW ARTICLES

Smiles rearrangement

J. F. Bunnett and R. E. Zahler, *Chem. Rev.*, 1951, **49**, 362–372.

W. E. Truce, E. N. Kreider, and W. W. Brand, *Org. Reactions*, 1970, **18**, 99.

Benzilic acid and related rearrangements

C. J. Collins and J. F. Eastham in *The Chemistry of the Carbonyl Group*, pp. 778–794, ed. Patai, Interscience (New York), 1966.

Favorskiĭ rearrangement

A. S. Kende, *Org. Reactions*, 1960, **11**, 261.

A. A. Akhram, T. K. Ustynyuk, and Yu. A. Titov, *Russian Chem. Rev.*, 1970, **39**, 732.

Ramberg–Bäcklund rearrangement

L. A. Paquette, *Accounts of Chem. Research*, 1968, **1**, 209.

Neber reaction

C. O'Brien, *Chem. Rev.*, 1964, **64**, 81.

Chapman rearrangement

J. W. Schulenberg and S. Archer, *Org. Reactions*, 1965, **14**, 1.

Chapman, Smiles, and Neber rearrangements

C. G. McCarty in *The Chemistry of the C—N Double Bond*, pp. 363–464, ed. Patai, Interscience (New York), 1970.

D. V. Banthorpe in *The Chemistry of the Amino Group*, pp. 585–668, ed. Patai, Interscience (New York), 1968.

7

Pericyclic Rearrangements I: Electrocyclic and Intramolecular Cycloaddition Reactions

In the classification of organic reaction mechanisms, two major and distinct types of reaction can be readily recognised, namely, those proceeding via a transition state where separation of charge is important (i.e., polar reactions) and those involving intermediate free-radical species. There exists a third class of reaction, diffuse in character, which contains many examples of molecular rearrangement. For such reactions, neither homolytic nor heterolytic bond-breaking or bond-making processes occur, although trends reminiscent of polar reactions are sometimes evident. Reactions of this type normally conform to the following pattern of behaviour.

(a) They proceed in a concerted fashion via a single transition state, i.e. bond-making and -breaking occur simultaneously and not sequentially.

(b) They are relatively insensitive to external catalytic influence (e.g. changes in solvent polarity) and to internal structural change, in striking contrast to the characteristics of reactions involving ionic intermediates.

(c) They show remarkable stereospecificity.

Because of the insensitivity of these reactions to the influences mentioned under (b), mechanistic study proved difficult. They were once categorised by Doering, in fact, as *no-mechanism reactions* as 'a humorous way of calling attention to the serious predicament organic chemists are in when the classical criteria of mechanism are not fulfilled'. This description was not intended to be taken literally, of course, since any reaction must have a mechanism in the sense that at least one transition state must intervene between reactant(s) and product(s). Reactions of this type are now less frivolously described as *pericyclic*, and in this and the following chapter pericyclic rearrangements will be discussed.

A coherent theoretical basis for the interpretation of pericyclic reactions was provided in 1965 by Woodward and Hoffmann when they enunciated the *principle of the conservation of orbital symmetry* [1]. It would be difficult to overstate the impact which this theory has exerted upon the practice of organic chemistry, and this important contribution to the interpretation of organic reactions was particularly mentioned in Woodward's Nobel Prize citation in 1968.

Pericyclic rearrangements may be subdivided into three types. In this chapter, electrocyclic rearrangements and intramolecular cycloaddition reactions will be considered. In the following chapter, sigmatropic rearrangements will be discussed.

Electrocyclic Rearrangements

The conversion of conjugated polyenes (1) to cyclopolyenes (2), by formation of a new σ-bond between the terminal carbon atoms of the conjugated chain and reorganisation of the π-electron system, or the reverse process, i.e. (2) → (1), are classified as electrocyclic reactions. Rearrangements of this type have been known for many years and are found to occur with almost total stereospecificity. As will become apparent in the sequel, the stereochemical course of such reactions is dependent not only upon the number of double bonds in the reactant but also upon the method by which reaction is initiated, either thermally or photochemically. Prior to the Woodward–Hoffmann rationale of 1965, it is not surprising that the stereospecificity found for such transformations presented a bewildering puzzle to organic chemists.

Δ or $h\nu$

(1) (2)

Δ CON

(3) (4)

Consider the experimental evidence. Thermally induced ring-opening of cyclobutenes (3) to buta-1,3-diene derivatives (4) proceeds readily at about 120° and in a highly stereospecific manner. The stereochemistry of the products from such reactions requires that the ring opens by a conrotatory (CON) movement of the groups attached to the saturated carbon atoms of the cyclobutene ring, as shown. Many reactions of the type (3) → (4) are known. For example, *cis*-3,4-dimethylcyclobutene (5) is converted on heating to *cis,trans*-hexa-2,4-diene (6) while the *trans*-isomer (7) yields the corresponding *trans, trans*-diene (8) [2].

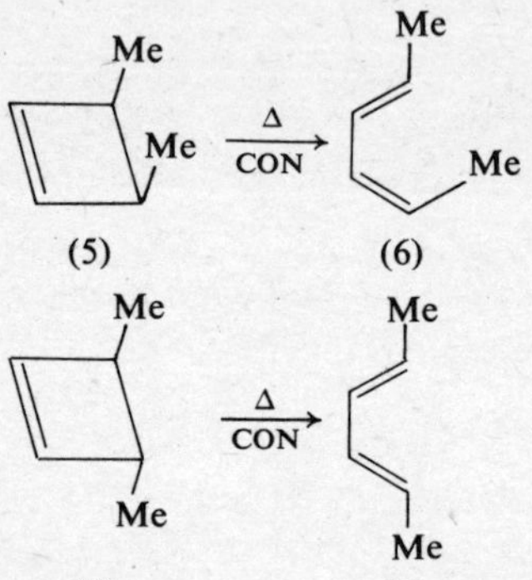

Me Me —Δ/CON→ Me Me

(5) (6)

Me Me —Δ/CON→ Me Me

(7) (8)

This thermal cyclobutene ⇄ butadiene interconversion should be thought of as an equilibrium process that involves forward and reverse reactions whose rates may differ substantially. For the simple substrates mentioned previously, the equilibrium position lies far on the side of the butadiene (4) which is thermodynamically much more stable than the strained cyclobutene precursor (3). When the cyclobutene moiety is part of a fused-ring system (9), however, the equilibrium position in the rearrangement (9) ⇄ (10) is strongly dependent upon the magnitude of the ring strain imposed upon the cycloalkadiene product (10).

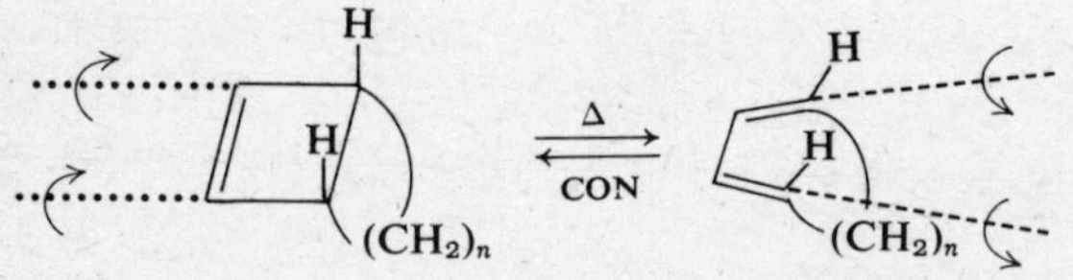

H H $(CH_2)_n$ ⇄ Δ/CON H H $(CH_2)_n$

(9) (10)

The conrotatory mechanism by which the σ-bond in the bicycloalkene (9) is broken under thermal initiation leads to the formation of a cycloalkadiene (10) in which the double bonds have a *cis*, *trans*-geometry. However, the angle strain occasioned by the presence of a *trans*-double bond in this cyclic structure becomes severe as the ring size is reduced below nine members. Consequently, whereas the bicyclodecene (9; $n = 6$) may be smoothly converted on heating to the diene (10; $n = 6$), the equilibrium position in the bicyclooctene $\rightleftarrows$ cyclooctadiene isomerisation (9 $\rightleftarrows$ 10; $n = 4$) lies far on the side of the bicyclic compound. Predictably, an intermediate situation obtains for the bicyclononene $\rightleftarrows$ cyclononadiene interconversion (9 $\rightleftarrows$ 10; $n = 5$) and substantial proportions of each product are present in the equilibrium mixture [3]. Irrespective of the position of equilibrium in these isomerisation reactions, however, it is important to note that these thermal interconversions proceed by a stereospecific conrotatory mechanism.

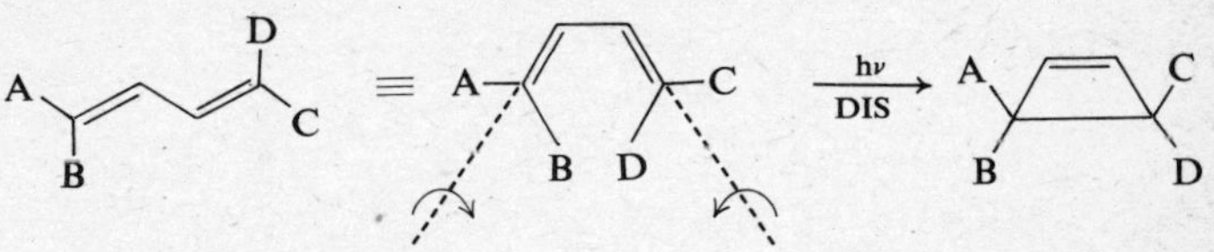

(11) (3)

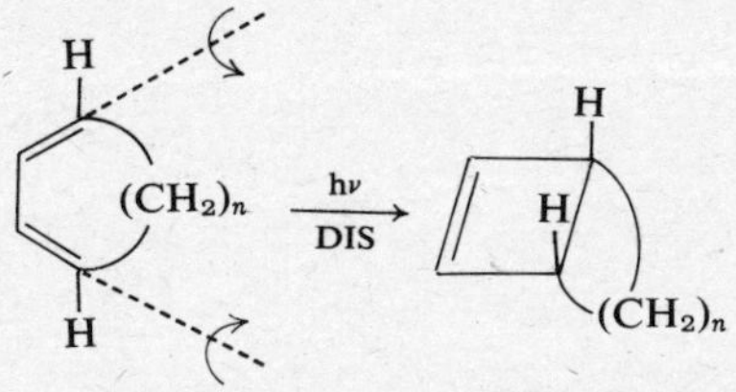

(12) (9)

Perfect stereospecificity has also been established for the photochemically induced interconversion of butadienes (11) and cyclobutenes (3). This reaction has important synthetic utility for the preparation of substituted cyclobutenes. The stereochemistry of the cyclised product requires that formation of the new σ-bond occurs by a disrotatory (DIS) process, as shown. A reversal of the situation found for the thermal reactions therefore prevails! Thus, *trans,trans*-hexa-2,4-diene (8), the product of thermal cleavage of *trans*-3,4-dimethylcyclobutene (7), undergoes photochemical cyclisation to *cis*-3,4-dimethylcyclobutene (5) [4]. Similarly, the *cis,cis*-cycloalkadienes (12; $n = 2$–5) cyclise readily upon ultraviolet irradiation to give *cis*-fused bicycloalkenes (9) by a disrotatory ring closure [5].

To summarise the experimental findings for the cyclobutene ⇄ butadiene interconversions discussed this far:

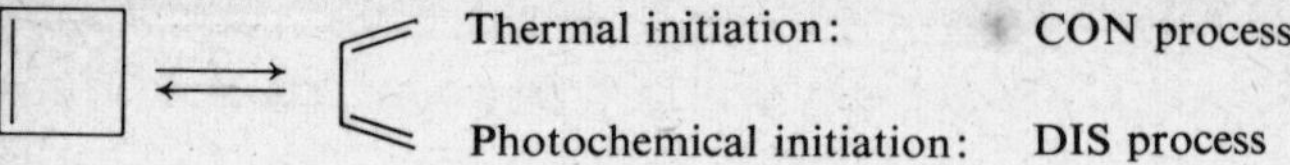

A related body of experimental evidence documents the stereospecificity associated with cyclohexa-1,3-diene ⇄ hexa-1,3,5-triene interconversions (13 ⇄ 14). However, as indicated below, a reversal of the situation established for the lower vinylogues is found:

Thermal initiation: DIS process

Photochemical initiation: CON process

(13) (14)

For example, the *cis*-fused bicyclononadiene (15) is isomerised on ultraviolet irradiation to the *cis,cis,trans*-triene (16), a conrotatory ring-opening occurring. This triene is unstable, due principally to angle strain, and cyclises on standing at room temperature. Under these thermal conditions, cyclisation occurs by a disrotatory process and the *trans*-fused bicyclic diene (17) results [6].

$h\nu$ CON — Δ DIS

(15) (16) (17)

The Woodward–Hoffmann rationale of the stereospecific nature of these electrocyclic reactions focuses attention upon the symmetry properties of the molecular orbitals that undergo change along the reaction coordinate. Consider the interconversion of cyclobutene and butadiene (1 ⇄ 2; $n = 0$). During the process, the four orbitals associated with the cyclobutene σ- and π-bonds which 'disappear' (i.e., σ, σ^*, π, and π^*) are transformed into the four π-type orbitals of the butadiene system (i.e. ψ_1, ψ_2, ψ_3, and ψ_4) or vice versa. The symmetries of these eight orbitals are indicated diagrammatically in Figure 1.

Figure 1

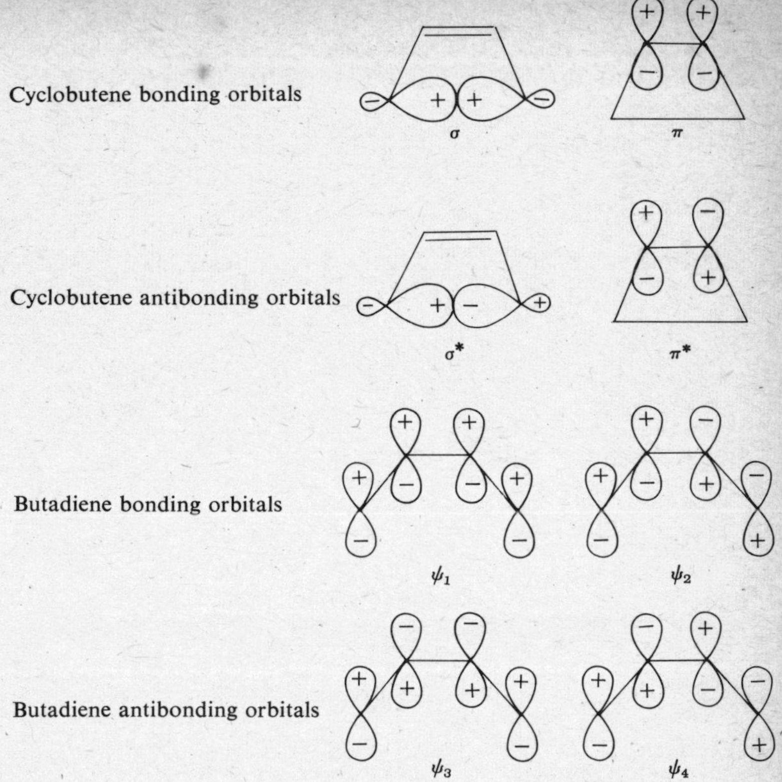

N.B. For clarity, the symmetries of these orbitals are indicated in terms of the signs of the wave-functions describing the component sp^3-hybrid or p atomic orbitals.

The key to an understanding of the energetics of the process hinges upon a recognition of which cyclobutene and butadiene orbitals correlate, depending upon the CON and DIS process by which the interconversion occurs. Consider a conrotatory opening of the cyclobutene σ-bonding orbital. This process must result in the production of a butadiene orbital with an odd number of nodes. Inspection of the symmetry properties of the orbitals which meet this requirement [i.e. ψ_2 (1 node) and ψ_4 (3 nodes); see Figure 1] reveals that, as the cyclobutene σ-bond is broken, the associated σ-orbital must 'mix' with the cyclobutene π^*-orbital to produce the ψ_2- and ψ_4-butadiene orbitals. A similar reasoning establishes that the cyclobutene σ^*- and π-orbitals are converted to the butadiene ψ_1- and ψ_3-orbitals by the

CON processes. These orbital interconversions are depicted in Figure 2.

Figure 2

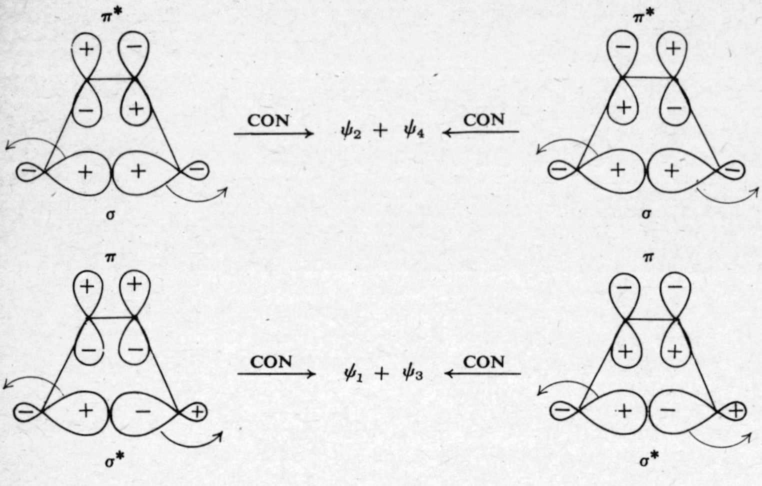

$$\xrightarrow{\text{CON}} \psi_2 + \psi_4 \xleftarrow{\text{CON}}$$

$$\xrightarrow{\text{CON}} \psi_1 + \psi_3 \xleftarrow{\text{CON}}$$

$$\text{Cyclobutene orbitals} \left\{ \begin{matrix} \sigma + \pi^* \rightleftarrows \psi_2 + \psi_4 \\ \sigma^* + \pi \rightleftarrows \psi_1 + \psi_3 \end{matrix} \right\} \text{Butadiene orbitals}$$

CON process

A similar analysis of the disrotatory interconversion leads to a different set of orbital correlations:

$$\text{Cyclobutene orbitals} \left\{ \begin{matrix} \sigma + \pi \rightleftarrows \psi_1 + \psi_3 \\ \sigma^* + \pi^* \rightleftarrows \psi_2 + \psi_4 \end{matrix} \right\} \text{Butadiene orbitals}$$

DIS process

These orbital interconversions for the CON and DIS processes may be depicted in orbital-correlation diagrams (Figure 3) in which the relative energy levels of the orbitals are indicated. Inspection of these diagrams reveals at once why the thermal interconversion of cyclobutene and butadiene proceeds by a CON mechanism. Such a process provides a low-energy pathway by which the electrons that are accommodated in the bonding orbitals of the reactant are relocated in the bonding orbitals of the product, i.e. cyclobutene (σ^2, π^2) $\rightleftarrows$ butadiene ($\psi_1{}^2$, $\psi_2{}^2$). Since the ground electronic states of reactant and

Figure 3

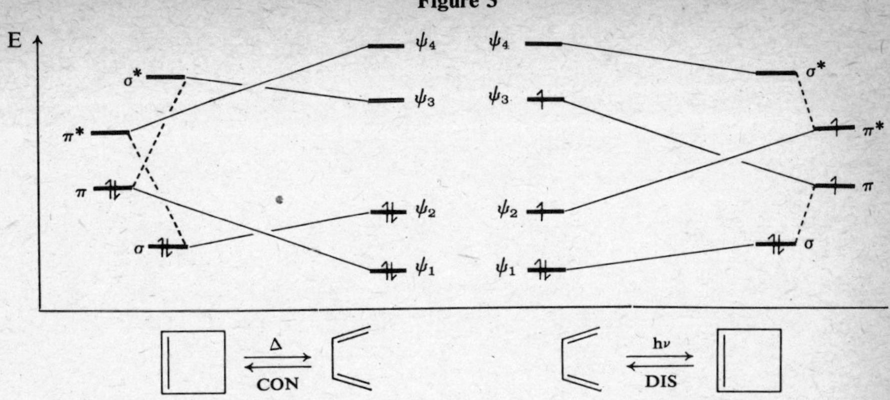

product are correlated by the CON mode, the interconversion is said to be 'symmetry-allowed'.

For a DIS interconversion, on the other hand, the ground electronic state of cyclobutene (σ^2, π^2) correlates with a doubly excited state of butadiene ($\psi_1{}^2$, $\psi_3{}^2$) while the ground electronic state of butadiene ($\psi_1{}^2$, $\psi_2{}^2$) correlates with a doubly excited state of cyclobutene (σ^2, π^{*2}). Interconversion of the *ground* electronic states of these molecules by a DIS mode is thus 'symmetry-forbidden' and a large energy barrier would be associated with such a change.

A different electronic situation obtains for the photochemical interconversion in which the reactant and product molecules are present in their first excited states i.e. (σ^2, π^1, π^{*1}) for cyclobutene and ($\psi_1{}^2$, $\psi_2{}^1$, $\psi_3{}^1$) for butadiene. A DIS interconversion of these excited states is 'symmetry-allowed' whereas the corresponding CON process is 'symmetry-forbidden' (see Figure 3).

The presence of substituent groups on the cyclobutene or butadiene skeleton of course alters the energy of the orbitals undergoing change, but does not alter their relative energy ordering. Thus, the CON process will always be favoured for thermal reactions and the DIS process for photochemical reactions, provided that these pathways are stereochemically feasible. In this connection, it is interesting to speculate that highly strained molecules such as bicyclo[2.1.0]pentene (18) and bicyclo[2.2.0]hexadiene (19; i.e., 'Dewar benzene') owe their existence and what little stability they possess to the fact that their conversion under thermal initiation to cyclopentadiene and benzene respectively would require a 'forbidden' DIS opening of their cyclobutene rings.

(18) (19)

Other electrocyclic rearrangements can be similarly analysed in terms of the symmetry properties of the interconverting orbitals. For the cyclohexadiene $\rightleftarrows$ hexatriene interconversion ($1 \rightleftarrows 2$; $n = 1$), for example, the following correlations can be derived.

$$\text{Cyclohexadiene orbitals} \left\{ \begin{array}{l} \sigma + \psi_2 + \psi_4 \rightleftarrows \psi_2 + \psi_4 + \psi_6 \\ \sigma^* + \psi_1 + \psi_3 \rightleftarrows \psi_1 + \psi_3 + \psi_5 \end{array} \right\} \text{Hexatriene orbitals}$$

CON process

$$\text{Cyclohexadiene orbitals} \left\{ \begin{array}{l} \sigma + \psi_1 + \psi_3 \rightleftarrows \psi_1 + \psi_3 + \psi_5 \\ \sigma^* + \psi_2 + \psi_4 \rightleftarrows \psi_2 + \psi_4 + \psi_6 \end{array} \right\} \text{Hexatriene orbitals}$$

DIS process

In this reaction, six orbitals of cyclohexadiene [i.e. the σ- and σ^*-orbitals associated with the bond which is broken and the ψ_1-, ψ_2-, ψ_3-, and ψ_4-orbitals (π-type) of the butadiene fragment] interchange with the six π-type hexatriene orbitals. Construction of an orbital-correlation diagram (cf. Figure 3) for each mode of interconversion reveals, then, that the DIS process is 'allowed' under thermal conditions and the CON process is 'allowed' under photochemical conditions, in harmony with the experimental findings.

From analyses of this nature, selection rules can be derived for electrocyclic rearrangements.

Thermal reactions. The DIS mode is 'allowed' for reactions in which $(4n + 2)$ electrons are accommodated in the orbitals undergoing change; i.e., for 2, 6, 10, etc. -electron systems (e.g., cyclohexadiene $\rightleftarrows$ hexatriene). The CON mode is 'allowed' for reactions in which $(4n)$ electrons are accommodated in the interconverting orbitals; i.e., 4, 8, 12 etc. -electron systems (e.g., cyclobutene $\rightleftarrows$ butadiene).

Photochemical reactions. Since the first excited states of the reactant molecules are involved, the thermal selection rules are reversed. Thus, for $(4n + 2)$-electron systems, the CON mode is 'allowed' and, for $(4n)$-electron systems, the DIS mode is 'allowed'.

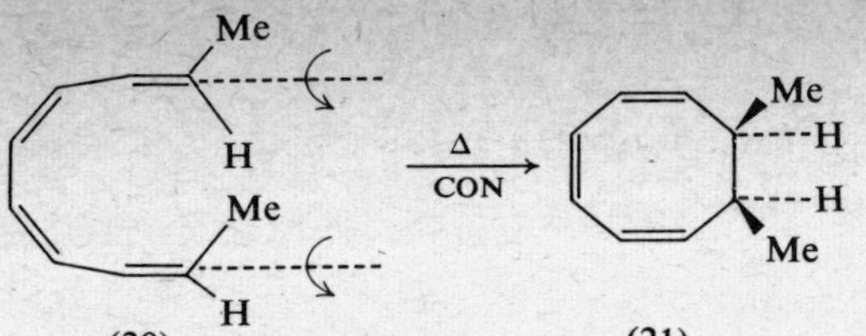

Me H Me H (20) $\xrightarrow[\text{CON}]{\Delta}$ Me H H Me (21)

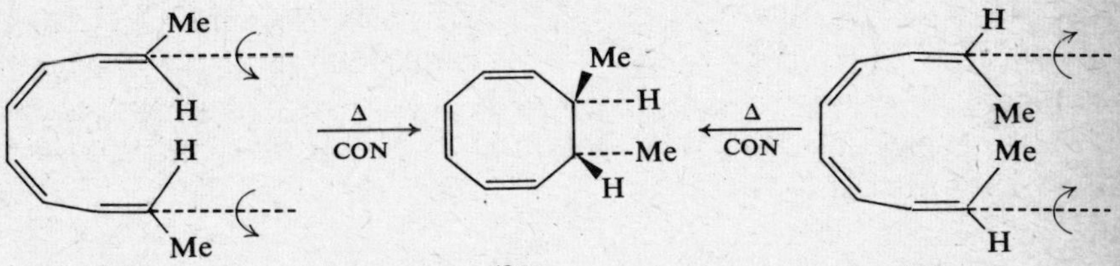

Me H H Me (22) $\xrightarrow[\text{CON}]{\Delta}$ Me H Me H (24) $\xleftarrow[\text{CON}]{\Delta}$ H Me Me H (23)

Since the derivation of these selection rules by Woodward and Hoffmann, their validity has been amply demonstrated. For example, thermally induced cyclisation of octatetraene (an 8π-electron system) should proceed by a CON mechanism. This prediction has been confirmed [7] by the transformations (20) → (21) and (22, 23) → (24).

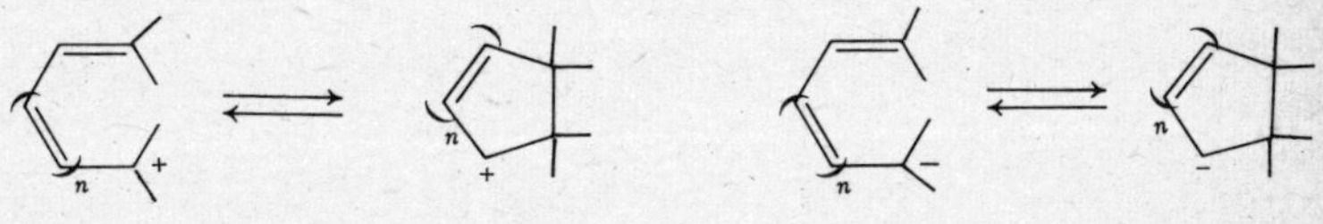

(25) ⇌ (26) (27) ⇌ (28)

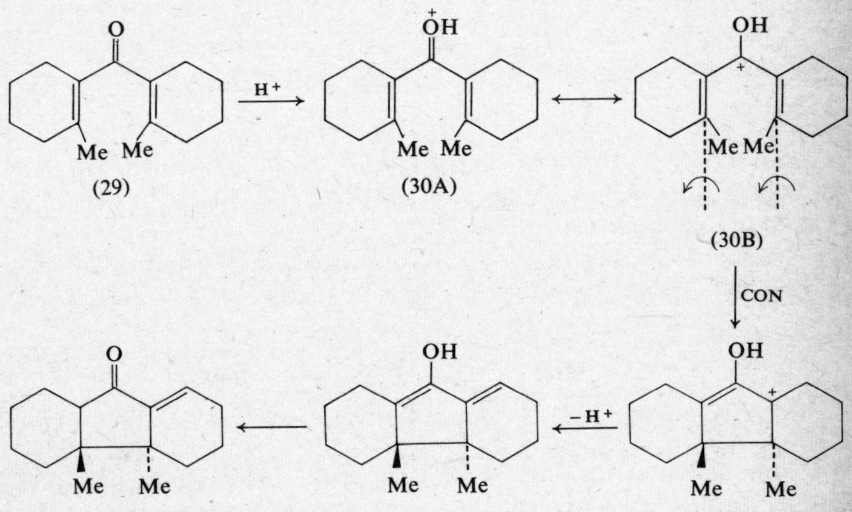

O Me Me (29) $\xrightarrow{H^+}$ $\overset{+}{O}H$ Me Me (30A) ⟷ OH + Me Me (30B)

CON ↓

O Me Me (32) ⟵ OH Me Me ⟵ $-H^+$ OH + Me Me (31)

The selection rules governing electrocyclic rearrangements should also be applicable to reactions, involving polyenyl cations and anions (i.e., 25 ⇄ 26 and 27 ⇄ 28). This has indeed been found to be so. For example, the di(methylcyclohexenyl) ketone (29) yields the tricyclic ketone (32) on treatment with phosphoric acid [8]. This compound no doubt arises by thermal cyclisation of the conjugate acid (30) of the ketone precursor, which can be regarded as a hydroxy-substituted pentadienyl cation (cf. 30A ⇄ 30B), a 4π-electron system. The *trans*-disposition of the methyl groups at the ring junction of the product (32) establishes that cyclisation (30 → 31) occurred by the predicted CON mode (see selection rules) followed by proton loss and enol → ketone tautomerism (31 → 32).

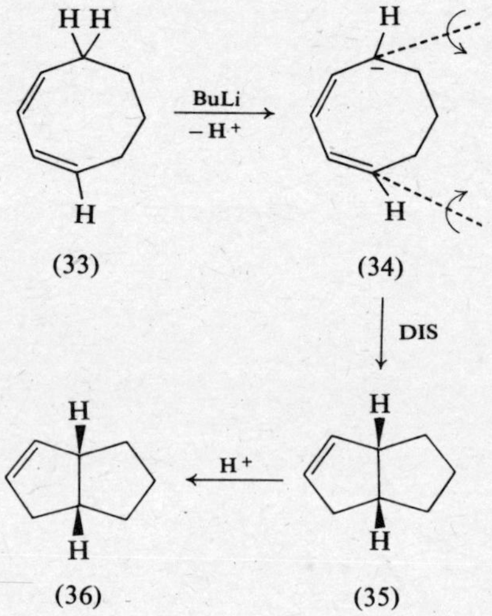

HH BuLi $-H^+$ H H H (33) (34) DIS H H H^+ H H (36) (35)

Similarly, the ready conversion [9] of cycloocta-1,3-diene (33) to the *cis*-fused bicyclo[3.3.0]octene (36) can be interpreted in terms of an 'allowed' DIS thermal cyclisation of the diene conjugate base (i.e. 34 → 35), a 6π-electron pentadienyl anion, generated by the reaction of the diene with butyl-lithium.

Intramolecular Cycloaddition Reactions

Concerted cycloaddition reactions of the type (37) + (38) → (39) between polyene systems are also subject to orbital-symmetry control. For reactions of this nature, Woodward and Hoffmann have derived selection rules [1] which may be summarised as follows.

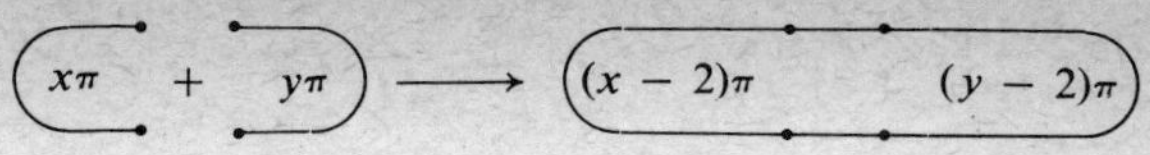

$x\pi$ + $y\pi$ ⟶ $(x-2)\pi$ $(y-2)\pi$

(37) (38) (39)

(a) Under thermal initiation, a concerted cycloaddition reaction of the type (37) + (38) → (39) is 'symmetry-allowed' where $x + y = 4n + 2$ (n = an integer) = 6, 10, 14, etc., and 'symmetry-forbidden' where $x + y = 4n$ = 4, 8, 12, etc. The well known Diels–Alder addition of an alkene (37; $x = 2$) to a conjugated diene (38; $y = 4$) to give a cyclohexene derivative is thus an example of a 'symmetry-allowed' $2\pi + 4\pi$ thermal cycloaddition reaction.

(b) Under photochemical initiation, the selection rules for concerted cycloaddition reactions are reversed. Reactions of the type (37) + (38) → (39) are 'symmetry-allowed' where $x + y = 4n$ and 'symmetry-forbidden' where $x + y = 4n + 2$. The photochemical dimerisation of alkenes (37, 38; $x = y = 2$) to give cyclobutane derivatives is thus an example of a 'symmetry-allowed' $2\pi + 2\pi$ photochemical cycloaddition reaction.

Irrespective of such orbital-symmetry restrictions, it is obvious that molecular rearrangements which involve a concerted intramolecular cycloaddition reaction, are stereochemically possible only for substrates in which the terminal atoms of the reacting π-systems can approach within bonding distance of each other. For this reason, the majority of the rearrangements of this type which have been discovered involve a $2\pi + 2\pi$ cycloaddition, i.e., the addition of a double bond to another double bond in the same molecule creating a cyclobutane ring. From the selection rules previously given, it will be seen that such reactions proceed under photochemical conditions.

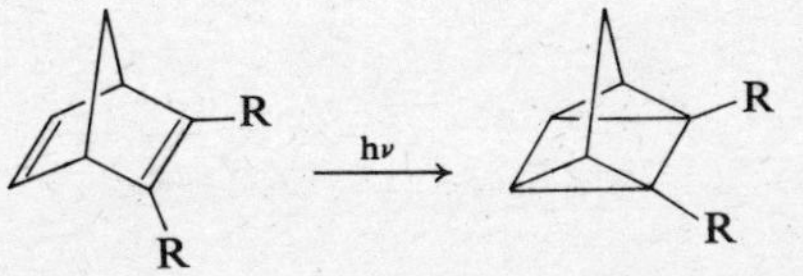

R R hν R R

(40) (41)

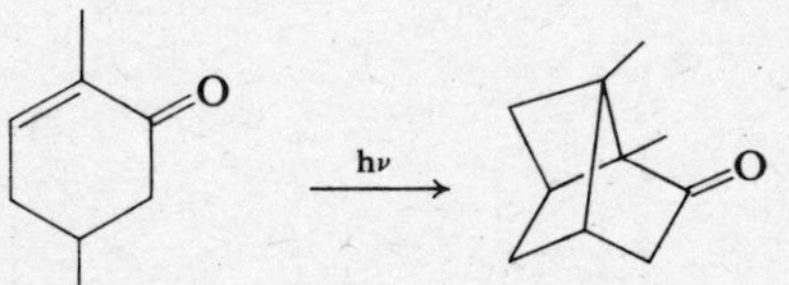

O hν O

(43)

(42)

(44) (45)

(46) (47)

Nevertheless, intramolecular cycloaddition reactions of the $2\pi + 2\pi$ type have proved invaluable to the organic chemist. Molecular frameworks of unusual construction, often highly strained, can be synthesised by reactions of this nature as illustrated by the transformations (40) → (41) (R = H [10], CO_2R [11]), (42) → (43) [12], (44) → (45) [13], and (46) → (47) [14]. In all of these rearrangements, it should be noted that the molecular geometry and/or conformational freedom of the substrates allow close approach of the reacting carbon-carbon double bonds. It is unnecessary, perhaps, to point out that the synthesis of the products from these reactions by alternative routes would pose sizeable problems for even the most inventive organic chemist!

References

1. R. B. Woodward and R. Hoffmann, *J. Amer. Chem. Soc.*, 1965, **87**, 395, 2046, 2511, 4388, 4389.
2. R. E. K. Winter, *Tetrahedron Letters*, 1965, 1207.
3. K. M. Shumate, P. N. Neuman, and G. J. Fonken, *J. Amer. Chem. Soc.*, 1965, **87**, 3996; P. Radlick and W. Fenical, *Tetrahedron Letters*, 1967, 4901.
4. R. Srinivasan, *J. Amer. Chem. Soc.*, 1968, **90**, 4498.
5. R. S. H. Liu, *J. Amer. Chem. Soc.*, 1967, **89**, 112; K. M. Shumate and G. J. Fonken, *J. Amer. Chem. Soc.*, 1966, **88**, 1073.
6. E. Vogel, W. Grimme, and E. Dinne, *Tetrahedron Letters*, 1965, 391.
7. R. Huisgen, A. Dahmen, and H. Huber, *J. Amer. Chem. Soc.*, 1967, **89**, 7130.
8. See R. B. Woodward and R. Hoffmann, *Angew. Chem. Internat. Edn.*, 1969, **8**, p. 806.

9. R. B. Bates and D. A. McCombs, *Tetrahedron Letters*, 1969, 977.
10. W. G. Dauben and R. L. Cargill, *Tetrahedron*, 1961, **15**, 197.
11. S. J. Cristol and R. L. Snell, *J. Amer. Chem. Soc.*, 1954, **76**, 5000; 1958, **80**, 1950.
12. G. Büchi and I. M. Goldman, *J. Amer. Chem. Soc.*, 1957, **79**, 4741.
13. R. Srinivasan, *J. Amer. Chem. Soc.*, 1963, **85**, 819.
14. R. C. Cookson, E. Crundwell, and J. Hudec, *Chem. and Ind.*, 1958, 1003.

Review Articles

R. B. Woodward in 'Aromaticity', *Chem. Soc. Special Publication* No. 21, pp. 217–249, 1967.

R. Hoffmann and R. B. Woodward, *Accounts of Chem. Research*, 1968, **1**, 17.

G. B. Gill, *Quart. Rev.*, 1968, **22**, 338.

S. I. Miller, *Adv. Phys. Org. Chem.*, 1968, **6**, 185.

J. J. Volmer and K. L. Servis, *J. Chem. Educ.*, 1968, **45**, 214.

L. Salem, *Chem. Brit.*, 1969, **5**, 449.

H. M. Frey and R. Walsh, *Chem. Rev.*, 1969, **69**, 103.

R. B. Woodward and R. Hoffmann, *Angew. Chem. Internat. Edn.*, 1969, **8**, 781.

R. B. Woodward and R. Hoffmann, *The Conservation of Orbital Symmetry*, Academic Press (New York), 1970.

J. B. Hendrickson, D. J. Cram, and G. S. Hammond, *Organic Chemistry*, Third Edn., Chapter 21, McGraw-Hill (New York), 1970.

D. C. Neckers, *Mechanistic Organic Photochemistry*, pp. 122–130, Reinhold (New York), 1967.

A rather different interpretation of pericyclic reactions from that presented in these Chapters is given by Dewar:

M. J. S. Dewar, *Angew. Chem. Internat. Edn.*, 1971, **10**, 761.

see also: C. L. Perrin, *Chem. Brit.*, 1972, **8**, 163.

8

Pericyclic Rearrangements II: Sigmatropic Reactions

As the description suggests, sigmatropic rearrangements involve the movement of a σ-bond from one position to another in a molecule, the bond shifting across one or two conducting π-electron systems whose double bonds are reorganised in the process. One of the simplest and best known examples of such a reaction is the Cope rearrangement ($1 \rightleftarrows 2$) (*vide infra*) in which a σ-bond linking two allylic fragments shifts from one pair of terminal allylic carbon atoms to the other.

A D B E (1) ⇄ A D B E (2)

In the organisation of the following discussion, the classification introduced [1] by Woodward and Hoffmann will be employed. In this system, sigmatropic rearrangements are classified by two numbers set in brackets thus $[i, j]$. These numbers indicate the particular atoms along the conducting conjugated chains to which each end of the migrating σ-bond becomes attached. The migration of a σ-bond across one chain only is therefore designated as a $[1, j]$ shift while $[i, j]$ rearrangements ($i, j \neq 1$) denote a σ-bond migration over two chains. This classification system is exemplified in Scheme 1. The Cope rearrangement ($1 \rightarrow 2$) is thus an example of a [3, 3] sigmatropic rearrangement. (*N.B.* The bracketed numbers denoting the order of the rearrangement should not be confused with those used in the systematic naming of a particular compound, e.g. by the I.U.P.A.C. rules.)

Scheme 1. Sigmatropic Shifts

R
|
C—C=C—C=C—C=C
1 2 3 4 5 6 7

[1, 3] → C=C—C(R)—C=C—C=C (1 2 3 4 5 6 7)

[1, 5] → C=C—C=C—C(R)—C=C (1 2 3 4 5 6 7)

[1, 7] → C=C—C=C—C=C—C(R) (1 2 3 4 5 6 7)

C—C=C—C=C (1 2 3 4 5)
|
C—C=C—C=C (1 2 3 4 5)

[1, 3] → C—C=C—C=C (1 2 3 4 5) bonded at C-1 to C-3 of C=C—C—C=C (1 2 3 4 5)

[3, 3] → C=C—C—C=C (1 2 3 4 5) bonded at C-3 to C-3 of C=C—C—C=C (1 2 3 4 5)

[3, 5] → C=C—C—C=C (1 2 3 4 5) bonded at C-3 to C-5 of C=C—C=C—C (1 2 3 4 5)

In sigmatropic rearrangements, the migration may involve movement of a bond to hydrogen, carbon, or hetero-atom, and the conducting π-system is not limited to a conjugated chain of carbon-carbon double bonds. The Claisen rearrangement (*vide infra*), for example, utilises a carbon-oxygen bond and examples of nitrogen analogues are also known.

[1, *j*] Sigmatropic Rearrangements

Many reactions have been found in which intramolecular migration of an atom or group occurs across a conjugated chain of double bonds. The following discussion will be restricted to rearrangements involving hydrogen or carbon migration, although hetero-atom shifts have also been recorded. These rearrangements occur in a stereospecific manner and follow different courses depending upon whether thermal or photochemical initiation is employed.

The stereospecific course of such transformations has been rationalised by Woodward and Hoffmann by application of orbital-symmetry concepts [1]. As for the electrocyclic reactions discussed in Chapter 7, all of the orbitals of the reactant molecule which undergo gross change can be correlated with those newly created in the product molecule, and an orbital-correlation diagram can be constructed for the reaction. However, the conclusions reached by such a laborious procedure can be derived more simply by consideration of the symmetries of the highest occupied orbitals of reactant and product, the 'frontier orbitals', which are involved in the reaction. In this chapter, therefore, this so-called 'frontier-orbital approach' will be described.

The transition state in a $[1, j]$ sigmatropic rearrangement can be constructed by consideration of the orbital symmetries of two interacting radical species, namely the migrating group and the molecular fragment over which it moves. It should be emphasised at this point, however, that discrete radical species do not intervene in concerted rearrangements, as will become apparent later. The molecular fragment over which the migrating group moves can be considered then

Figure 1. Nonbonding Orbitals of Polyenyl Radicals

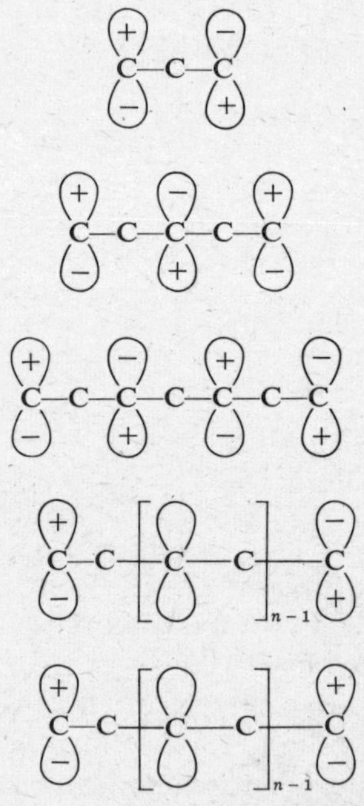

Allyl [(C=C)C·]

Pentadienyl $[(C{=}C)_2C\cdot]$

Heptatrienyl $[(C{=}C)_3C\cdot]$

Polyenyl $[(C{=}C)_nC\cdot]$
(n = odd integer)

Polyenyl $[(C{=}C)_nC\cdot]$
(n = even integer)

as a polyenyl radical [i.e. $(C{=}C)_nC\cdot$] and, for the species of this type, the odd electron is located in a nonbonding orbital (i.e. the highest occupied orbital) in the ground (unexcited) state. The symmetries of these nonbonding polyenyl orbitals are indicated in Figure 1. It will be noted that the nodal properties of these orbitals (i.e. nodes at every second carbon atom) are such that the wave-function signs of the orbital lobes on either side of the carbon framework are reversed at every third carbon atom along the chain.

Migration of hydrogen. Consider the sigmatropic migration of a hydrogen atom. Such rearrangements are known to proceed in a concerted intramolecular fashion. The bond between the hydrogen atom and the migration terminus must therefore begin to form before the bond to the migration origin is severed. In other words, the migrating hydrogen atom must be partially bound both to the migration origin and terminus simultaneously. Since the hydrogen atom can only use its spherically symmetrical 1*s*-orbital for bonding purposes, it is obvious that simultaneous bonding to two atoms of a polyenyl fragment is possibly only if the orbital lobes of the latter which are used have the same sign.

Scheme 2

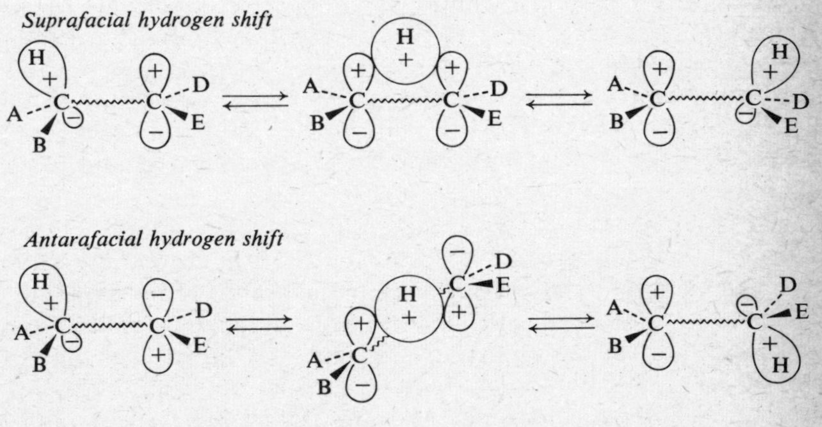

Suprafacial hydrogen shift

Antarafacial hydrogen shift

With this proviso in mind, a hydrogen shift can then occur, in principle, in two stereochemically distinct ways that involve either the migration of the atom over one face of the polyenyl moiety or the migration from one face of the polyenyl chain to the opposite face. These rearrangement processes are designated *suprafacial* and *antara-*

facial respectively and are depicted in Scheme 2. It will be noted that suprafacial and antarafacial shifts produce different stereochemical consequences at the migration terminus. If two different groups are attached to this carbon atom in the precursor (i.e. D $\neq$ E; see Scheme 2), a suprafacial hydrogen shift to this centre produces a product different from that which results from an antarafacial shift. The same stereochemistry will result at the migration origin, however, irrespective of the migration mode involved.

The selection rules for $[1, j]$ sigmatropic shift of a hydrogen atom under thermal conditions therefore follow from the polyenyl orbital symmetries depicted in Figure 1.

Thermal rearrangements

(a) A $[1, j]$ suprafacial hydrogen migration is 'symmetry-allowed' where $j = 4n + 1$ (n = an integer), e.g., a [1, 5] shift.

(b) A $[1, j]$ antarafacial hydrogen migration is 'symmetry-allowed' where $j = 4n - 1$, e.g., a [1, 7] shift. Note that an antarafacial [1, 3] hydrogen shift, which is 'symmetry-allowed', is sterically impossible since the migrating hydrogen atom cannot be simultaneously bonded to opposite faces of a three-carbon allylic group.

Photochemical rearrangements

Under photochemical conditions, the reacting molecule is in an electronically excited state. As in the case of an electrocyclic rearrangements discussed in Chapter 7, this leads to a reversal of the selection rules which apply to the corresponding thermal reactions. Thus, photochemically induced $[1, j]$ hydrogen shifts are 'symmetry-allowed' for a suprafacial migration where $j = 4n - 1$, and for an antarafacial migration where $j = 4n + 1$.

A large number of experimental results have now been obtained which fully substantiate these selection rules for hydrogen migration. For example, the stereospecific hydrogen migrations found [2] in the thermal and photochemical isomerisations of cycloheptatrienes (see Scheme 3) can be accounted for in terms of 'symmetry-allowed' suprafacial [1,5] shifts in the thermal reactions and suprafacial [1,7] shifts in the photochemical reactions. Antarafacial migrations in such seven-membered ring systems, of course, are sterically impossible.

The preference for [1,5] over [1,3] or [1,7] shifts in the thermal isomerisation of polyene systems has been demonstrated [3] by Roth who investigated the reaction of cycloocta-1,3,5-triene (3) bearing deuterium labels at C_7 and C_8. For this compound, a series of [1,3] or [1,7] hydrogen shifts would result in a statistical distribution of deuterium

Scheme 3

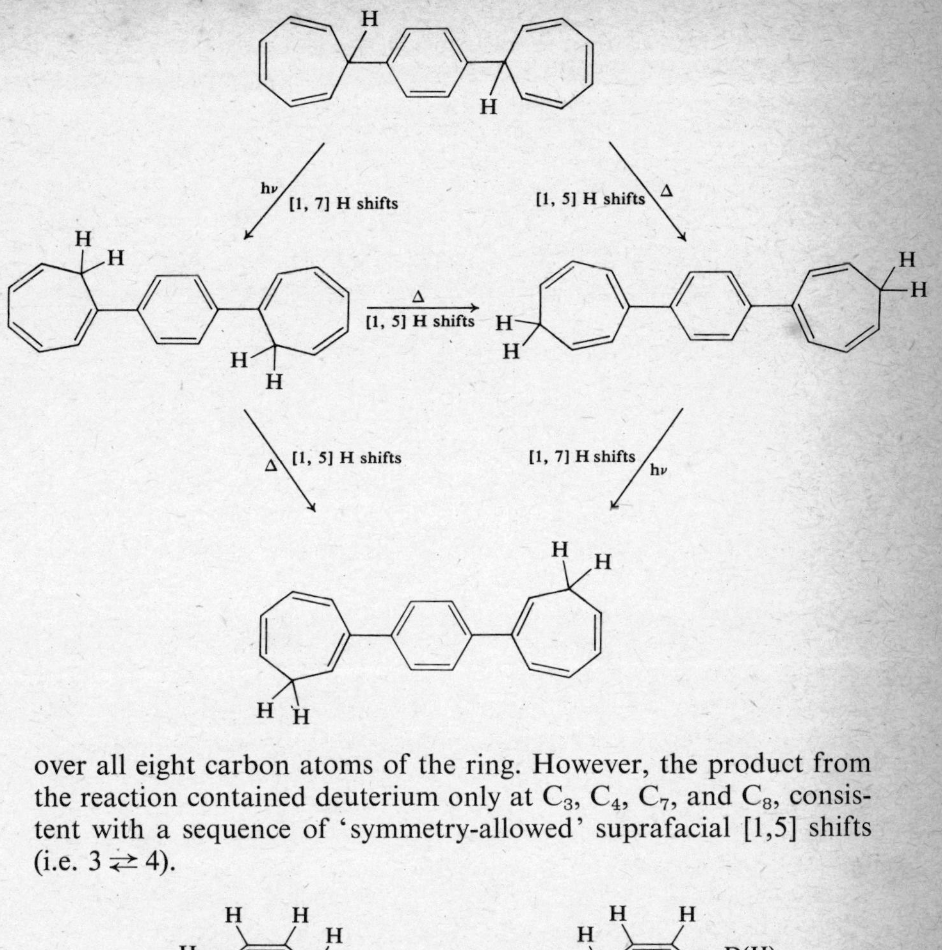

over all eight carbon atoms of the ring. However, the product from the reaction contained deuterium only at C_3, C_4, C_7, and C_8, consistent with a sequence of 'symmetry-allowed' suprafacial [1,5] shifts (i.e. 3 ⇄ 4).

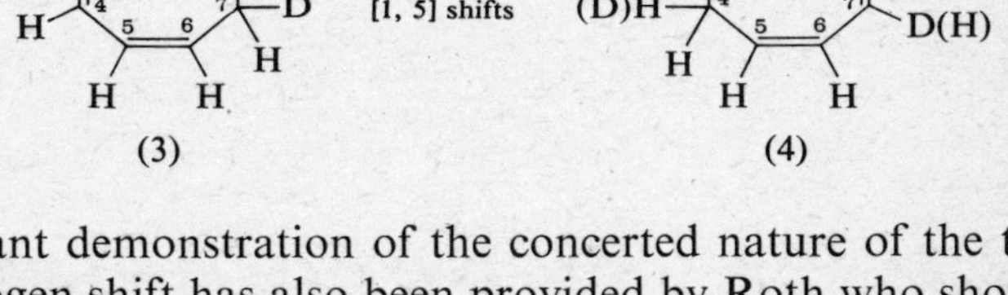

(3) (4)

An elegant demonstration of the concerted nature of the thermal [1,5] hydrogen shift has also been provided by Roth who showed [4] that the diene (5) is stereospecifically converted to a mixture of the isomeric dienes (6 and 7) on heating. The precursor can adopt two rotameric conformations (5A and 5B) in which suprafacial hydrogen

Scheme 4

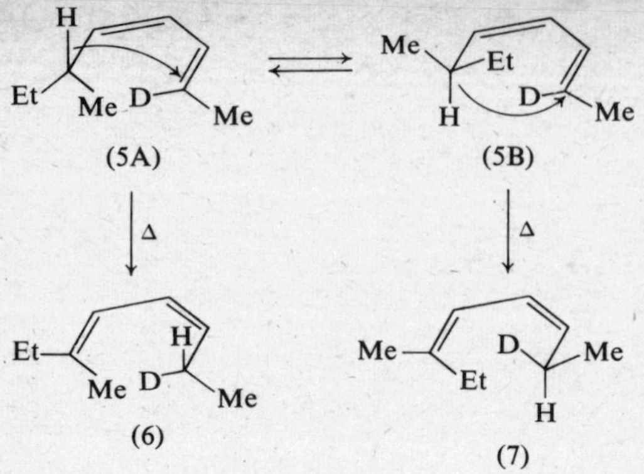

(5A) ⇌ (5B)

(6) (7)

migration can occur. The stereochemistry of the products of the isomerisation establishes that the reaction must proceed as indicated in Scheme 4 by concerted [1,5] suprafacial shifts.

Migration of carbon. At first sight, it might appear that sigmatropic migration of a carbon atom over a polyenyl fragment would be governed by the same selection rules as those established for hydrogen migration. However, there is one important difference between the migrating capacities of hydrogen and carbon atoms. It was previously pointed out that a hydrogen atom can utilise only its spherically symmetrical 1*s*-orbital in the transition state for migration with the attendant symmetry restrictions which this places upon the rearrangement pathways open. A carbon atom, on the other hand, may use orbitals which possess *p*-character (i.e. either 'pure' *p* or sp^x-hybrid variety). Orbitals of this type are characterised by a node at the atomic nucleus such that a carbon atom, unlike hydrogen, may interact simultaneously with two orbital lobes of opposite sign during a sigmatropic migration across a polyenyl moiety. The course of such a rearrangement process is depicted in Scheme 5.

Scheme 5

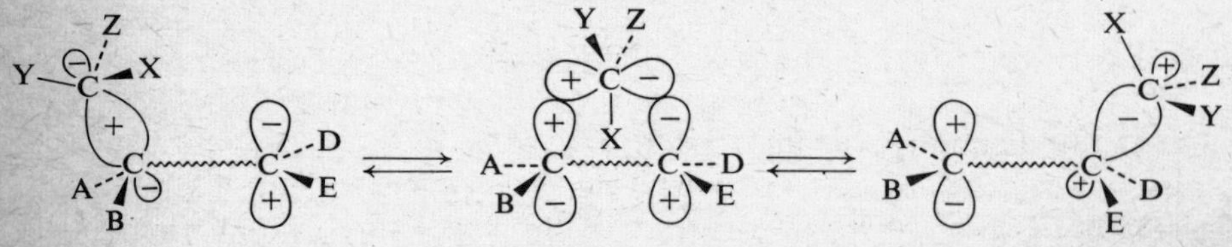

Careful consideration of the steric course of such a concerted carbon atom shift reveals that the migration must occur with inversion of configuration at the migrating atom. Consequently, if the migrating carbon atom is asymmetrically substituted (i.e. $X \neq Y \neq Z$; see Scheme 5), the arrangement of groups around this atom in the product is stereochemically different from the arrangement found for the precursor. From the foregoing, it follows that thermal $[1, j]$ suprafacial carbon shifts become 'symmetry-allowed' where $j = 4n - 1$, provided that the rearrangement proceeds with inversion at the migrating atom. This requirement is beautifully illustrated [5] by the stereospecific conversion of the deuterium-labelled bicyclo[3.2.0]-heptenyl acetate (8) to the norbornenyl acetate (9) at about 300°. The stereochemistry of the product in this reaction shows that the carbon atom which executes a suprafacial [1,3] shift does so with inversion of configuration (see Scheme 5).

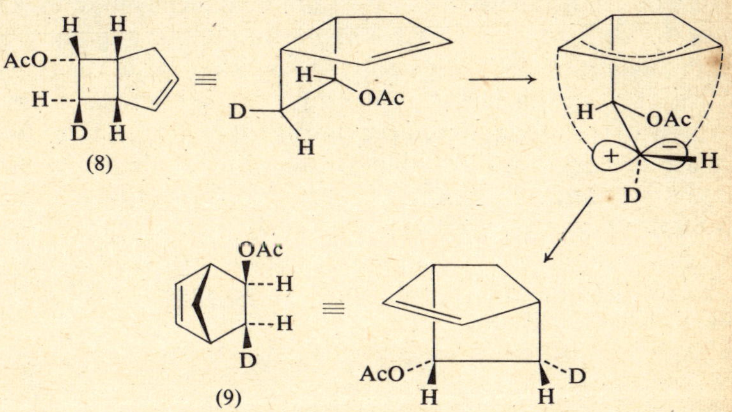

The thermal isomerisation of the optically active ester of Feist's acid (10) has been studied in this connection [6a]. The two isomeric esters formed in this rearrangement have been shown to have the structures (11) and (12). Their formation can be interpreted as involving a concerted [1,3] shift of the cyclopropane C_2—C_3 bond (see 10) which can occur in two discrete ways. For each, the structure of the product requires that the reaction, if concerted, must occur with inversion of configuration at the migrating atom (C_2 or C_3), in accordance with the selection rules. More recent work [6b], however, leads to the conclusion that methylenecyclopropane rearrangements of this type may proceed via chiral diradical intermediates (non-planar trimethylenemethanes).

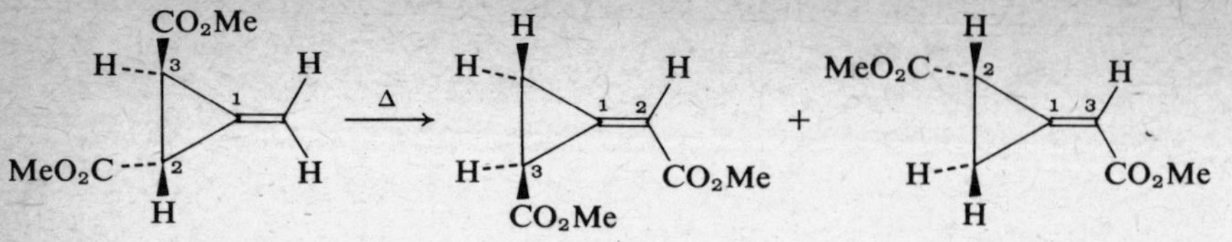

(10) (11) (12)

Suprafacial $[1,j]$ carbon shifts where $j = 4n + 1$ and antarafacial $[1,j]$ carbon shifts where $j = 4n - 1$ are 'symmetry-allowed' only if they occur with retention of configuration at the migrating atom. Since the migrating atom uses only one orbital lobe in the transition state for the rearrangement, the symmetry requirements are thus identical with those established previously for hydrogen migration.

Concerted sigmatropic rearrangements within ionic species are also commonly encountered in organic chemistry and may be treated from an orbital-symmetry viewpoint. For such processes, the selection rules can be readily derived by consideration of the relative orbital symmetries of a migrating radical and a polyene radical cation or anion as the case may be.

Scheme 6

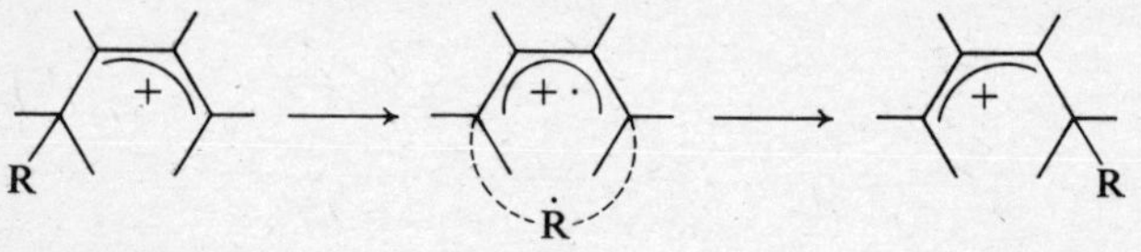

(13) (14) (15)

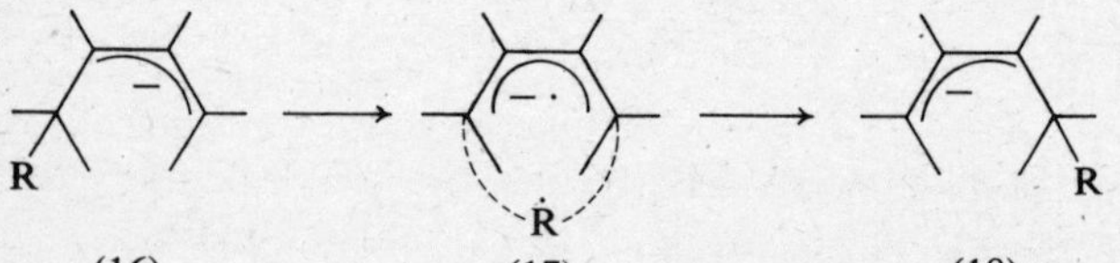

(16) (17) (18)

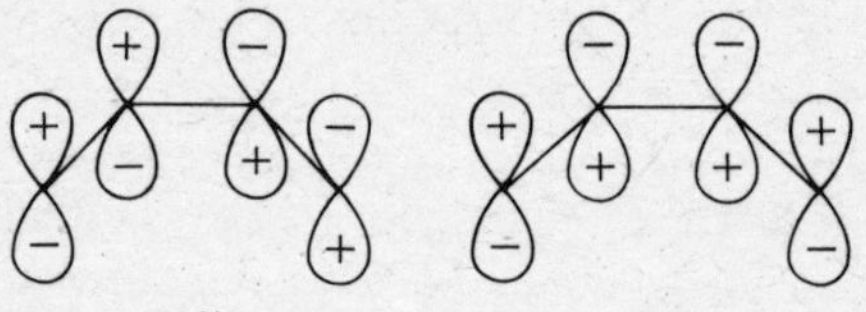

(19) (20)

For example, consider the concerted sigmatropic migration of a group R across a cationic or anionic allylic group (i.e. 13 → 15 or 16 → 18 respectively; see Scheme 6). The transition state for these rearrangements can be constructed in terms of the 'symmetry-allowed' orbital interactions between the migrating radical R and either a butadiene radical cation (14) or radical anion (17). These latter species may be simply regarded as the charged radicals resulting either by removal from or by addition of a π-electron to butadiene (i.e. three or five π-electron systems respectively). The symmetries of the highest occupied level of these species accordingly are different. For the radical cation, the odd electron is accommodated in the ψ_2 butadiene orbital of symmetry (19) while the odd electron in the radical anion is accommodated in the ψ_3 butadiene orbital of symmetry (20).

From inspection of the symmetry relationship between the lobes at the termini of these orbitals (19 and 20) and bearing in mind the orbitals which the migrating groups can utilise (*vide supra*), the following selection rules for thermally induced [1,4] sigmatropic shift of hydrogen or carbon can be deduced.

(a) A [1,4] suprafacial shift of hydrogen is 'symmetry-allowed' for the anionic (16; R = H), but not for the cationic substrate (13; R = H).

(b) A [1,4] suprafacial shift of carbon is 'symmetry-allowed' for the cation (13; R = CR'_3) provided that the migration occurs with inversion of configuration at the migrating atom.

(c) A [1,4] suprafacial shift of carbon is 'symmetry-allowed' for the anion (16; R = CR'_3) provided that the migration occurs with retention of configuration at the migrating atom.

(d) The conclusions stated under (a), (b), and (c) are reversed for antarafacial migrations, although rearrangements of this type would be sterically difficult if not impossible.

The predictive power of orbital-symmetry correlations of this type has been demonstrated in many cases. For example, debromination of the bicyclo[3.1.0]hexanone derivative (21) with zinc in refluxing dioxan yields [7] the product (24). The course of this transformation can be interpreted as proceeding via the dipolar intermediate (22), a substituted allylic cation. The stereochemistry of the product (24) establishes that the [1,4] suprafacial migration of C_3 from C_2 to C_4 (i.e. 22 → 23) must occur as predicted with inversion of configuration at C_3.

The selection rules for [1,*j*] sigmatropic rearrangements within ionic species are summarised below.

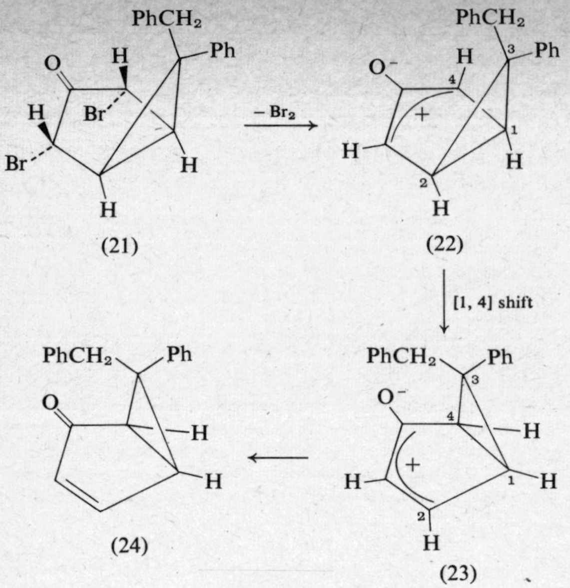

PhCH2 Ph O H H Br Br H H

(21)

$-Br_2$ →

PhCH2 3 Ph O⁻ H 4 + 1 H H 2 H

(22)

[1, 4] shift

PhCH2 Ph 3 O⁻ 4 H + H 1 H 2 H

(23)

←

PhCH2 Ph O H H

(24)

Cationic species

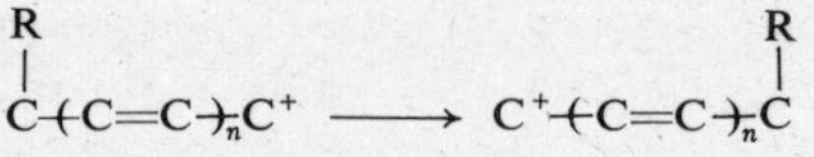

$$\underset{\textstyle(25)}{\overset{\textstyle R}{\overset{|}{C}}\!-\!(C{=}C)_n\!-\!C^+} \longrightarrow \underset{\textstyle(26)}{C^+\!-\!(C{=}C)_n\!-\!\overset{\textstyle R}{\overset{|}{C}}}$$

'Symmetry-allowed' processes where n = odd integer, e.g. [1,4] or [1,8] shifts:

antarafacial hydrogen migration
suprafacial carbon migration with inversion
antarafacial carbon migration with retention.

'Symmetry-allowed' processes where n = even integer, e.g. [1,6] or [1.10] shifts:

suprafacial hydrogen migration
suprafacial carbon migration with retention
antarafacial carbon migration with inversion.

Anionic species

The selection rules given above for cationic species are exactly reversed.

It will be noted that the special case of a [1,2] hydrogen shift in a carbonium ion (i.e. 25 → 26; $n = 0$; R = H), which for steric reasons could only occur in a suprafacial manner, is 'symmetry-allowed'. Such migrations are very commonly observed for carbonium ions (e.g. $CH_3CH_2\overset{+}{C}H_2 \rightarrow CH_3\overset{+}{C}HCH_3$). The corresponding carbon shift (i.e. 25 → 26; $n = 0$; R = CXYZ) is required to proceed with retention of configuration at the migrating atom in harmony with experimental findings (see Chapter 2).

A *concerted* [1,2] shift of hydrogen to an anionic centre (i.e. 27 → 28; $n = 0$; R = H) is 'symmetry-forbidden' and it is not surprising

$$\underset{(27)}{\overset{\displaystyle R}{\overset{|}{C}}\!\!-\!\!(C{=}C)_n\!-\!C^-} \longrightarrow \underset{(28)}{C^-\!-\!(C{=}C)_n\!-\!\overset{\displaystyle R}{\overset{|}{C}}}$$

that this rearrangement has not been found. Concerted [1,2] carbon shifts to a negative centre are predicted to occur with inversion of configuration at the migrating atom. The Stevens rearrangement, which involves a [1,2] shift of carbon from nitrogen to a carbanion centre, however, has been found to occur mainly with retention of configuration. The mechanism of this reaction, however, may involve radical or ion-pair intermediates and the selection rules for concerted migrations would not apply for such stepwise processes (see Chapter 5).

[*i*,*j*] Sigmatropic Rearrangements

Rearrangements which involve the migration of a σ-bond over two conjugated π-electron systems may also be analysed from orbital-symmetry concepts. The transition state for such reactions can be visualised as resulting from the 'symmetry-allowed' interaction of two mesomeric polyenyl radicals. Perhaps it should be emphasised again, however, that concerted sigmatropic rearrangements do not proceed via free-radical intermediates, i.e., homolytic bond cleavage does not occur and electrons remain paired throughout the structural reorganisation. Analysis of the rearrangement transition state in terms of interacting radical species merely serves as a pointer to the orbital interactions which are permitted on symmetry grounds.

Reference to the symmetry relationships between the terminal lobes of the polyenyl nonbonding orbitals (see Figure 1) shows that, under thermal conditions, an [*i*,*j*] sigmatropic rearrangement is 'symmetry-allowed' where $(i + j) = (4n + 2)$, i.e., $(i + j) = 6, 10, 14$, etc. Thus, [3,3] shifts are 'allowed' (see Scheme 7) whereas [3,5] shifts are

Scheme 7. [3, 3] Sigmatropic Rearrangements

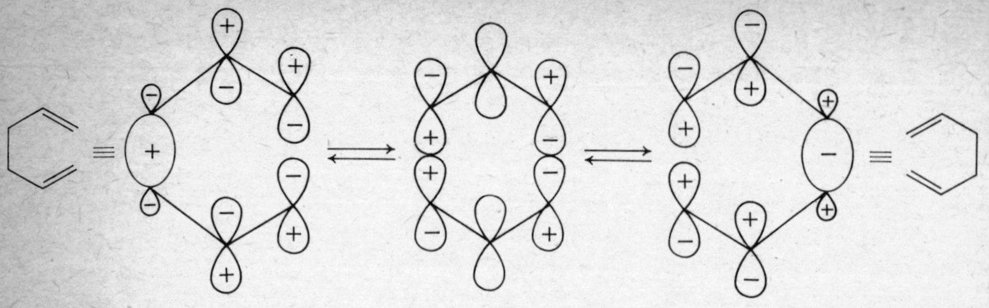

'forbidden'. In photochemical reactions, the rearranging molecule is electronically excited and the selection rules are reversed in consequence, e.g. [3,5] shifts are 'allowed' and [3,3] shifts are 'forbidden'.

The most commonly encountered examples of $[i, j]$ sigmatropic rearrangements in organic chemistry are the [3,3] variety which are 'symmetry-allowed' under thermal conditions. The best known examples are the Cope and Claisen rearrangements and these will be discussed in turn.

Cope rearrangement. In 1940, Cope discovered [8a] that the unsaturated ester (29) is converted to its isomer (30) merely on heating to 150°. In a series of elegant investigations reported [8b] in subsequent years, Cope and his collaborators soon established that the minimum structural requirement for the rearrangement is the presence of a hexa-1,5-diene system, and that an equilibrium between the isomeric species (e.g. 31 ⇄ 32) is established in the reaction.

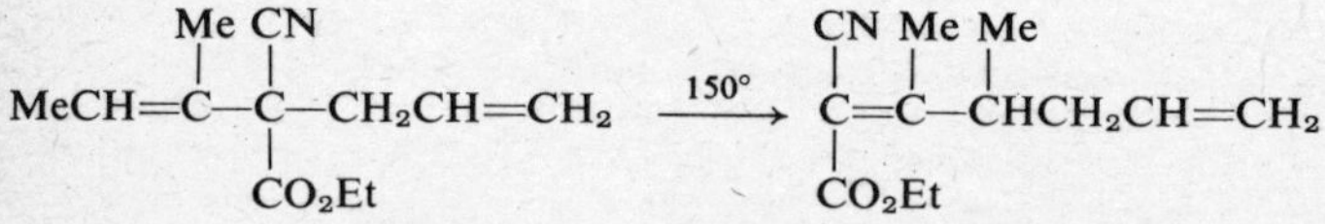

$$MeCH{=}C(Me){-}C(CN)(CO_2Et){-}CH_2CH{=}CH_2 \xrightarrow{150^\circ} (CO_2Et)(CN)C{=}C(Me){-}CH(Me)CH_2CH{=}CH_2$$

(29) (30)

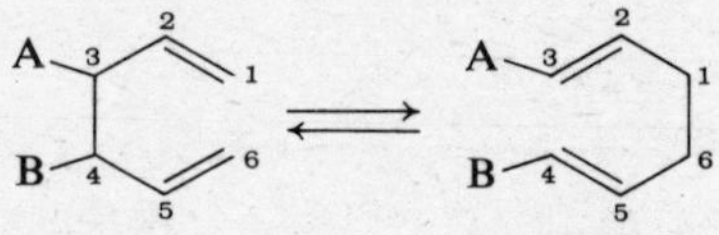

(31) ⇄ (32)

For example, either hepta-1,5-diene (32; A = Me, B = H) or its isomer (31; A = Me, B = H) is slowly converted to an equilibrium mixture of both dienes at 300°. For systems in which one compound is thermodynamically much more stable than its isomer, of course, the former will largely predominate in the equilibrium mixture. (*N.B.* For such a process of the type A $\rightleftarrows$ B, a free-energy difference of only 4 kJ/mole between the interconverting species will produce an equilibrium mixture containing about 80 per cent of the thermodynamically more stable isomer at 25 °C). It is understandable, therefore, that there is observed a ready rearrangement of hexa-1,5-diene derivatives, bearing groups attached to the saturated carbon atoms (i.e., C_3 and C_4 in 31) that can conjugate with the newly created double bonds in the product, thereby providing resonance stabilisation (e.g. 31 $\rightarrow$ 32; A, B = Ph, CN, CO_2R).

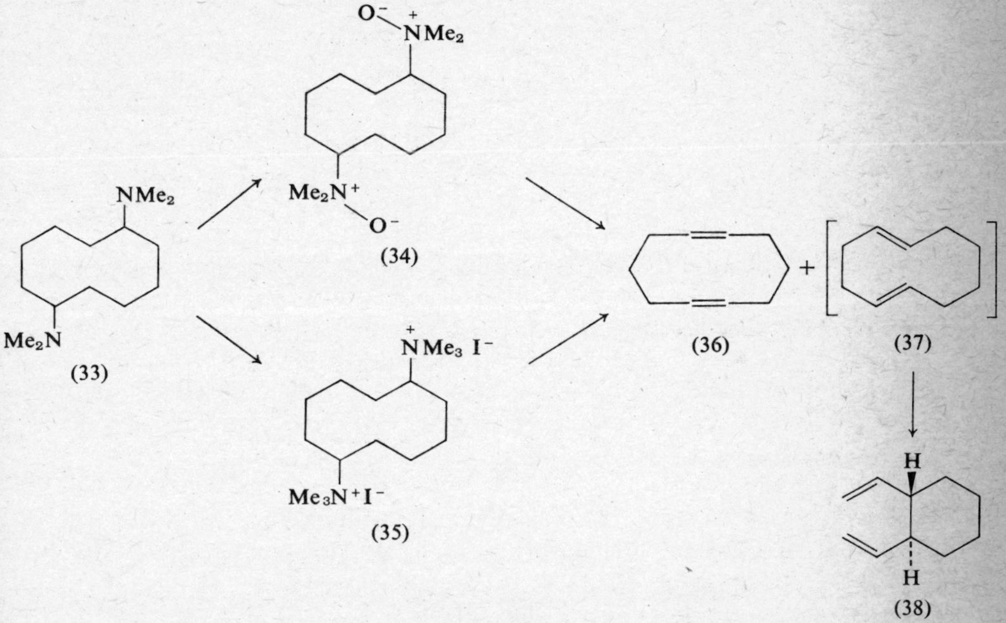

(33) (34) (35) (36) (37) (38)

Relief of steric strain also provides a driving force for the Cope rearrangement. For example, the diamine (33) can be converted to cyclodeca-1,6-diene (36; *cis*/*trans* isomeric mixture) via either the bis(*N*-oxide) (34) or the bisquaternary salt (35) using, respectively, the Cope or Hofmann elimination procedures. These reactions produce substantial amounts of *trans*-1,2-divinylcyclohexane (38), however. It may be assumed that these elimination reactions produce *trans*,*trans*-cyclodeca-1,5-diene (37) in addition to the 1,6-diene isomers (36).

Naturally, rearrangement of this strained 1,5-diene to a strainless cyclohexane derivative occurs readily [9].

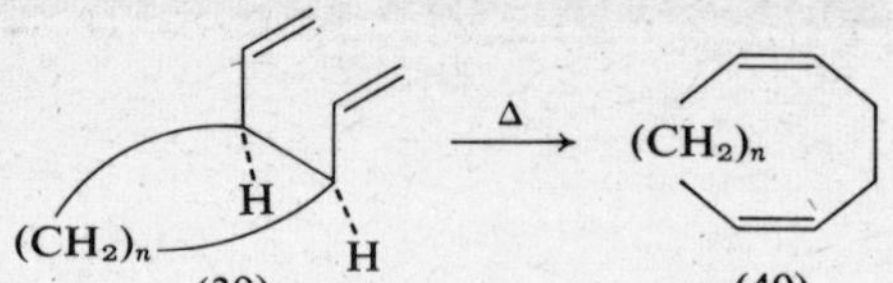

(39) (40)

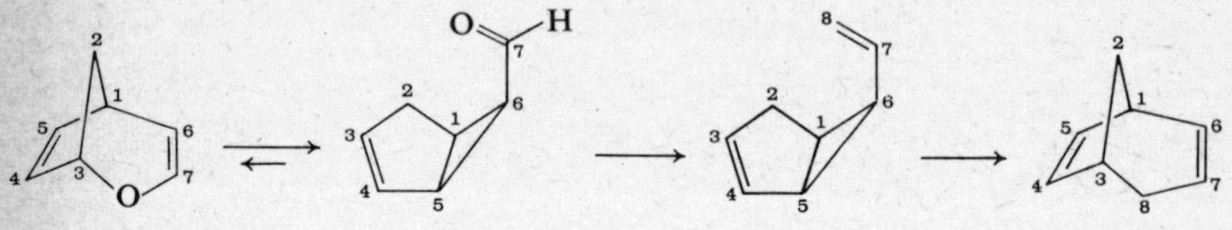

(41) (42) (43) (44)

Similarly, *cis*-1,2-divinylcyclobutane (39; $n = 2$) is converted at 120° to cycloocta-1,5-diene (40; $n = 2$) in high yield [10]. The relief of angle and eclipsing strain afforded by disruption of the cyclobutane ring provides the driving force for the reaction. The even more highly strained cyclopropane analogue (39; $n = 1$), in fact, rearranges so readily that it cannot be isolated; conversion to cyclohepta-1,4-diene (40; $n = 1$) occurs under the conditions of its formation [11] (e.g., by a Hofmann elimination reaction). This ready rearrangement of *cis*-1,2-divinylcyclopropanes can be put to use for the synthesis of compounds which are difficult to obtain by other routes. For example, the aldehyde (42), which is readily prepared from bicyclo[2.2.1]heptadiene via its mono-epoxide, undergoes a Wittig reaction with methylenetriphenylphosphorane ($CH_2{=}PPh_3$) to give the diene (43). This derivative readily rearranges thermally (half-life of one day at 25°) to give bicyclo[3.2.1]octadiene (44) by a Cope rearrangement [12]. Bicyclo[3.2.1]octadienes substituted at C_8 are also readily obtainable through use of a substituted phosphorane ($RR'C{=}PPh_3$) in the reaction sequence [13].

Parenthetically, it is interesting to note that the aldehyde (42) itself is unstable and rearranges reversibly [14] to the bridged bicyclic ether (41). The equilibrium mixture contains a preponderance of the cyclopropyl compound, and this interconversion represents a Claisen-type oxygen analogue of the all-carbon Cope rearrangement. The parent compound (45) itself shows similar behaviour [15] and, at room temperature, exists in equilibrium with its valence tautomer (46). The unsaturated ether (47) is so unstable to Cope-type rearrangement, in fact, that it is converted to the isomeric aldehyde (48) below room temperature [16]. Nitrogen analogues of the Cope rearrangement of *cis*-divinylcyclopropanes have also been found to provide easy synthetic routes to dihydroazepines [17] (e.g. 49 → 50) and dihydrodiazepines [18] (e.g. 51 → 52).

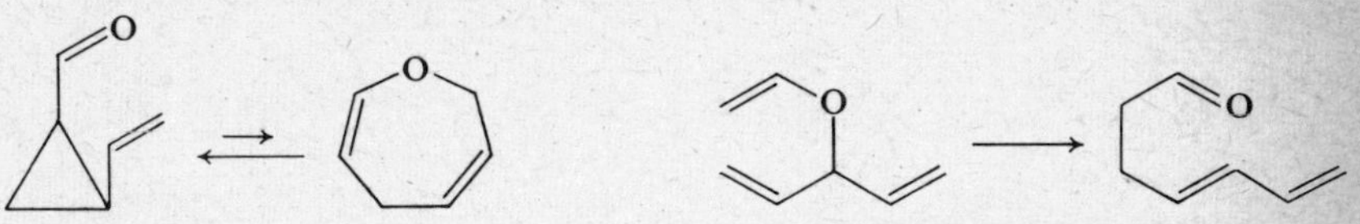

(45) ⇄ (46)   (47) → (48)

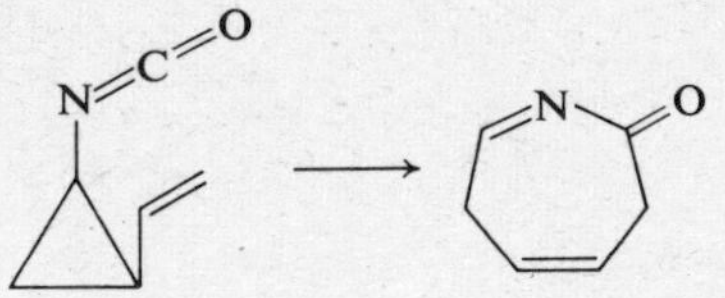

(49) → (50)

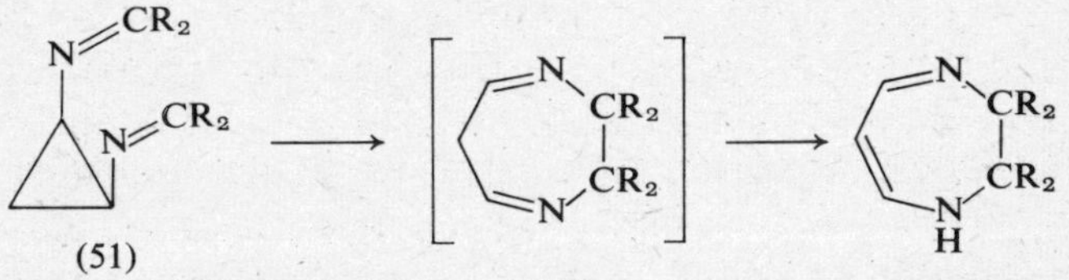

(51) → [ ] → (52)

Cope rearrangement of hexa-1,5-diene itself would produce an identical molecule (i.e. 31 ⇄ 32; A = B = H). The occurrence of self-perpetuating reactions of this type, which are known as degenerate rearrangements, has actually been demonstrated by proton magnetic resonance spectroscopy. For example, the p.m.r. spectrum of bicyclo[5.1.0]octadiene at room temperature is in accord with the structure (53A). At 180°, however, the corresponding spectrum shows that the molecule rapidly undergoes a reversible rearrangement to produce an identical structure (i.e. 53A ⇄ 53B). At this elevated temperature, the rate of interconversion is greater than 1000 times per second [19]. This phenomenon is a special case of valence tautomerism and a molecule exhibiting such behaviour is said to have a fluxional structure. Valence tautomeric structures need not be identical, however. For example, bicyclo[4.2.0]octa-2,4-diene (54) and cycloocta-1,3,5-triene (55) are stable molecules at room temperature but, at about 100°, they exist in tautomeric equilibrium with each other, the monocyclic compound predominating [20] [15% (54) ⇄ 85% (55)].

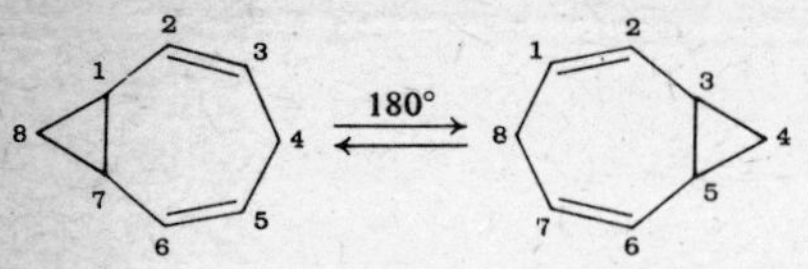

(53A) (53B) 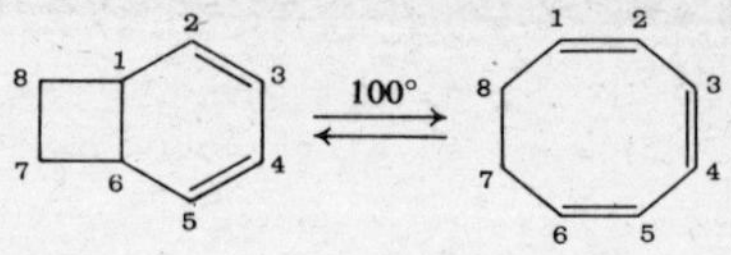 (54) (55)

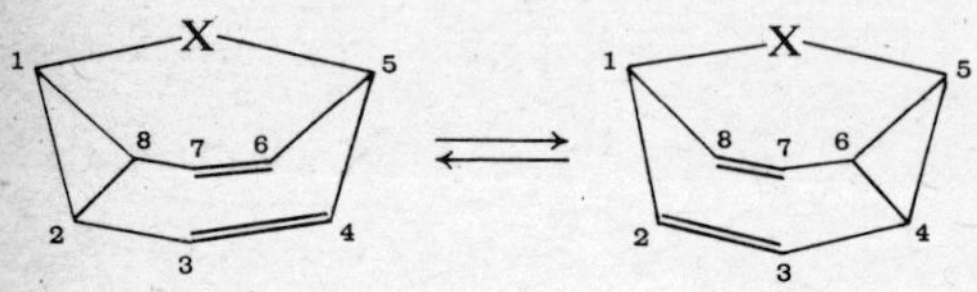

(56A) (56B) 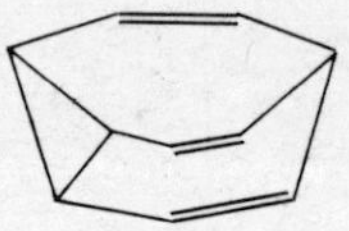 (57)

Many fluxional organic molecules have been found which interconvert with themselves at room temperature or below, e.g., (56A) ⇄ (56B) (X = CH_2 [21] and X = CO [22]). The most spectacular example of this behaviour is provided by bullvalene (57), which incorporates an *all cis*-1,2,3-trivinylcyclopropane structure. Cope rearrangement involving any pair of double bonds may occur generating an identical molecule. In consequence, the cyclopropane ring may be formed by any three adjacent carbon atoms at any instant of time and there are therefore 10!/3 (i.e. more than a million) identical tautomeric forms possible for this bizarre molecule. Bullvalene has been synthesised [23]. At −25°, its p.m.r. spectrum is in accord with a static non-fluxional structure (57). At 100°, however, the spectrum consists of a single resonance, indicating time-averaged equivalence of all ten protons present in the molecule. At this temperature, therefore, the compound is rapidly interchanging its structure among its vast number of identical tautomeric forms.

The geometrical requirements and the stereochemical consequences of the Cope rearrangement have been investigated. From consideration of the examples previously discussed, it will be apparent that, in every case, the double bonds of the rearranging hexa-1,5-diene system are located in close juxtaposition. That this steric relationship is actually an essential requirement for the rearrangement is illustrated by the behaviour of the isomeric 1,2-divinylcyclobutanes [11]. It has been mentioned previously that the *cis*-isomer (39; n = 2) rearranges readily. The double bonds in the *trans*-isomer (58), however, are unable to approach closely together. This compound does not undergo a Cope rearrangement on thermolysis but instead fragments into two molecules of butadiene. Both *cis*- and *trans*-1,2-divinylcyclopropane rearrange thermally to cyclohepta-1,4-diene [10, 11]. Whereas the *cis*-isomer rearranges spontaneously at room temperature, however, the *trans*-isomer (59) reacts only at 200°. It seems probable that this reaction involves isomerisation of the *trans*- to the *cis*-isomer via a diradical intermediate and thence to the cycloheptadiene as indicated (59 → 60 → 61).

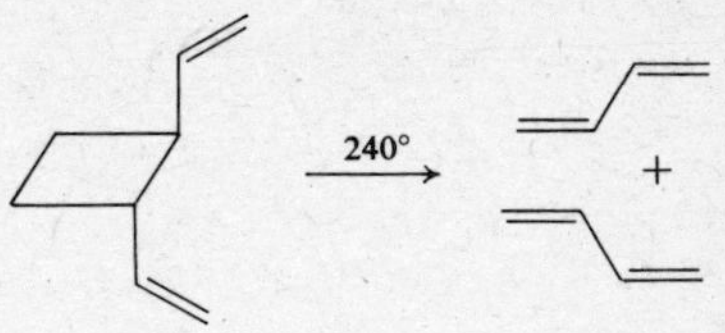

240°

+

(58)

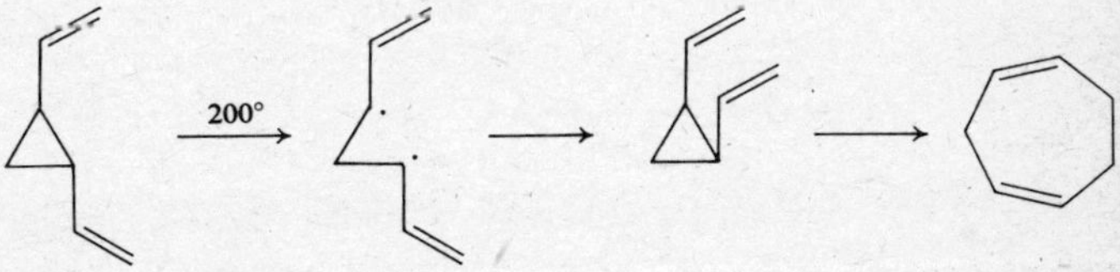

200°

(59) (60) (61)

In principle, the transition state in the Cope rearrangement can be pictured as a quasi-chair or quasi-boat structure as shown in Scheme 8. For substrates which are free to adopt either stereochemistry (i.e., non-cyclic hexa-1,5-dienes), Doering and Roth have shown [24] that

Scheme 8. Cope Rearrangement Transition States

Quasi-Chair transition state

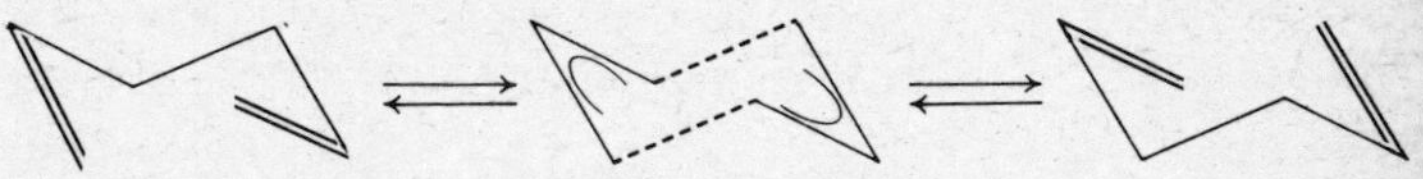

Quasi-Boat transition state

the quasi-chair geometry is definitely preferred. Thus, Cope rearrangement of *meso*-3,4-dimethylhexa-1,5-diene (62) at 225° yields octa-2,6-diene consisting of 99·7 per cent of the *cis-trans*-isomer (64). The racemic form of the precursor (65A, 65B), on the other hand, rearranges at 180° to give a mixture of 90 per cent *trans,trans*- and 9 per cent *cis,cis*-octa-2,6-diene (67 and 69 respectively).

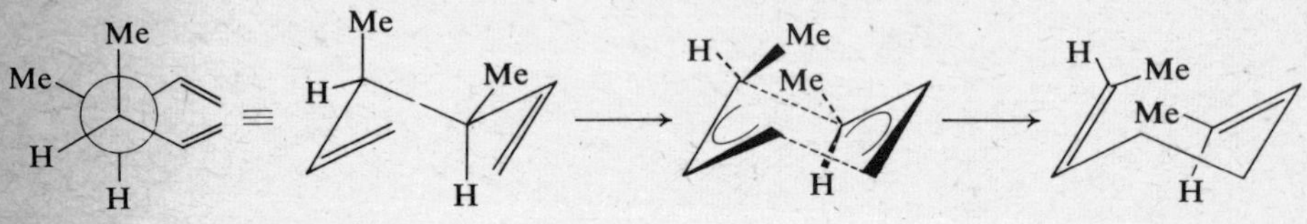

(62) (63) (64)

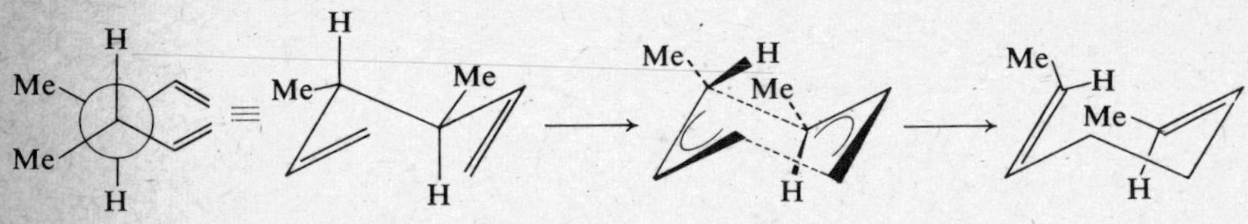

(65A) (66) (67)

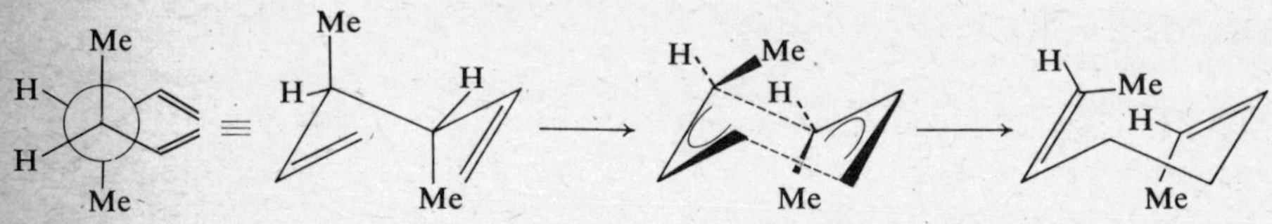

(65B) (68) (69)

These results establish clearly that the rearrangements proceed almost exclusively through chair-like transition states, (63) in the case of the *meso*-diene and either (66) or (68) in the case of the racemic diene. In the latter rearrangement, the conformation (66) is preferred over (68), in which the methyl groups are sterically more crowded, and the *trans,trans*-diene (67) accordingly predominates in the product. Although the chair-like transition state is therefore preferred over the boat-like for those substrates given a choice of either, the energy difference between these two arrangements cannot be very large. Thus, compounds which can rearrange only through a quasi-boat transition state may still undergo reaction under mild conditions (e.g., *cis*-1,2-divinylcyclopropanes; *vide supra*).

A fascinating and deceptively simple example of a Cope rearrangement which must proceed via a boat-like transition state has been described [25] by Woodward and Katz. The epimeric alchols (70; R = H; R′ = OH and R = OH, R′ = H), derived from cyclopentadiene dimer, rearrange stereospecifically at about 140° to give equilibrium mixtures of the starting material and a single rearranged alcohol in each case. For the α-epimer, the product contains approximately

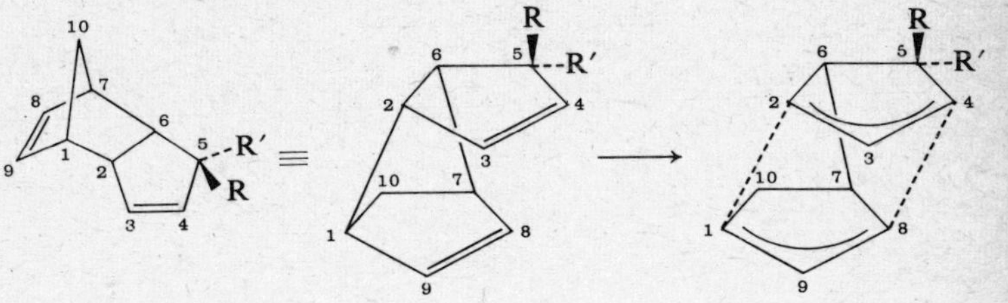

(70)

(71)

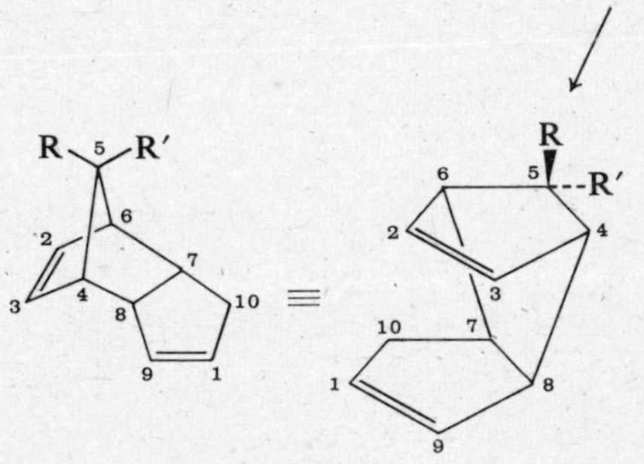

(72)

equal amounts of each alcohol [i.e., 50% (70) ⇄ 50% (72); R = H, R′ = OH]. The β-epimer (70; R = OH, R′ = H), however, is less stable than its rearrangement product (72; R = OH, R′ = H), which is stabilised by hydrogen-bonding between the hydroxyl group and the C_2—C_3 double bond, and the equilibrium mixture contains about 90 per cent of the latter alcohol. These rearrangements must proceed through a transition state of the boat type (71). In the structural reorganisation, the five-membered rings linked by the C_6—C_7 bond swivel relative to each other with a concomitant [3,3] sigmatropic shift of the C_1—C_2 bond to a C_4—C_8 location.

Claisen rearrangement. Allyl aryl ethers (73) rearrange smoothly to *o*-allylphenols (74) on heating at about 200° either in absence of solvent or in solution in an inert high-boiling solvent such as *N*,*N*-dimethylaniline. This rearrangement was first discovered [26] in 1912 by Claisen and has since been found to be generally applicable to ethers of the type (73) in which at least one *ortho*-position of the aryl ring is unsubstituted. High yields of product (often > 80 per cent) are usually obtainable and the reaction has proved useful in organic synthesis, e.g., in the two-step preparation [27] of *o*-eugenol (76) from guaiacol (75).

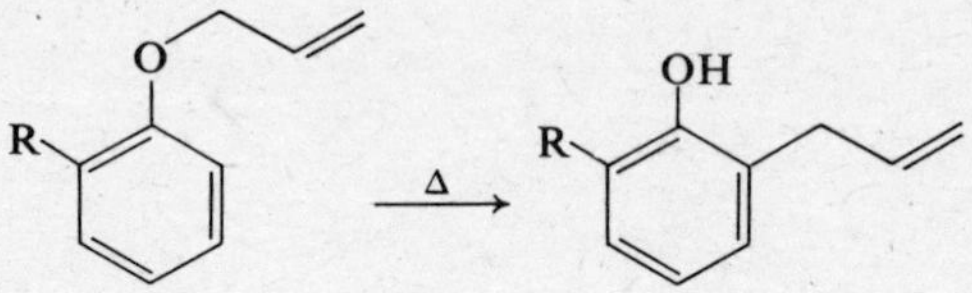

(73) $\xrightarrow{\Delta}$ (74)

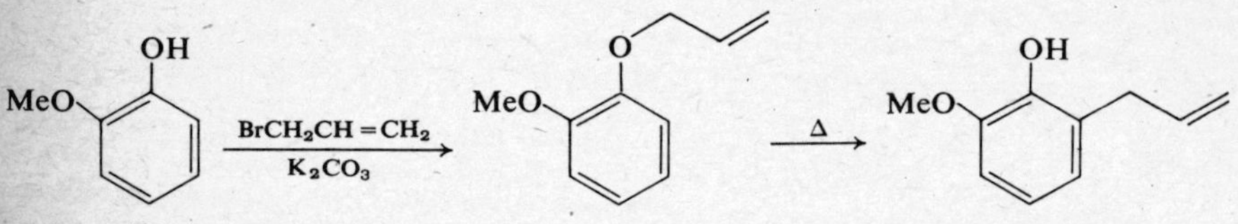

(75) $\xrightarrow[K_2CO_3]{BrCH_2CH=CH_2}$ $\xrightarrow{\Delta}$ (76)

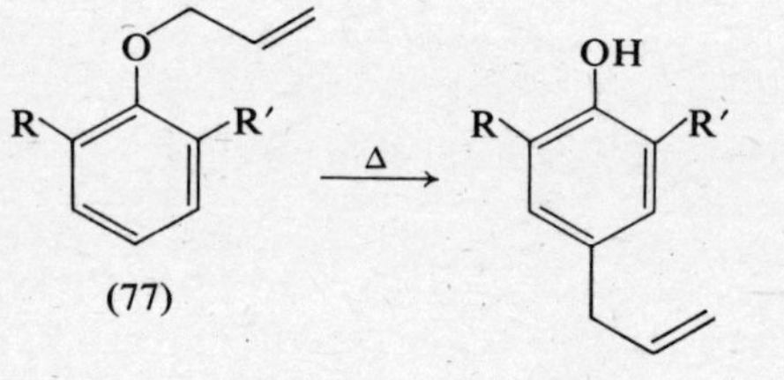

(77) $\xrightarrow{\Delta}$ (78)

Aryl allyl ethers (77), in which both *ortho*-positions of the aromatic ring are substituted, also undergo thermal rearrangement in which the allyl group migrates to the *para*-position giving a *p*-allylphenol (78). Such products of Claisen *para*-rearrangement are also sometimes observed in the thermolysis of *ortho*-monosubstituted ethers (73) although usually the product of *ortho*-rearrangement (74) greatly predominates.

Mechanistic studies [28] of the Claisen rearrangement have established that both the *ortho*- and *para*-rearrangements are first-order, intramolecular processes proceeding through highly-ordered cyclic transition states. The first clues to the mechanism of the reaction came from the observation that it is the γ-carbon atom of the allyloxy group which becomes attached to the *ortho*-position of the benzene

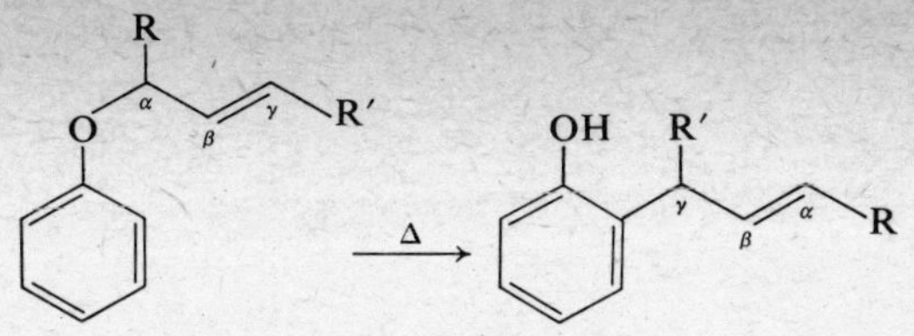

(79) (80)

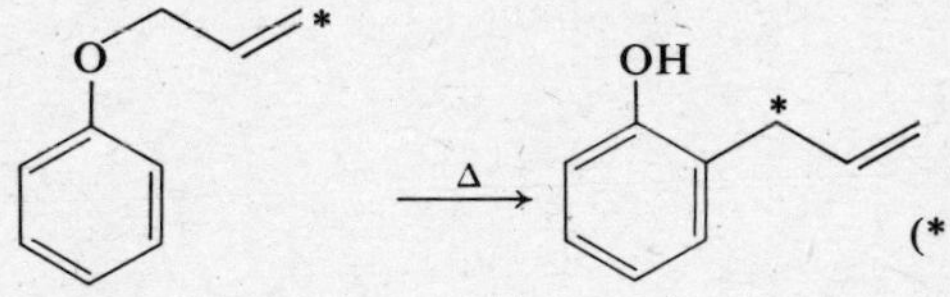

(81) (82)

(* denotes ^{14}C)

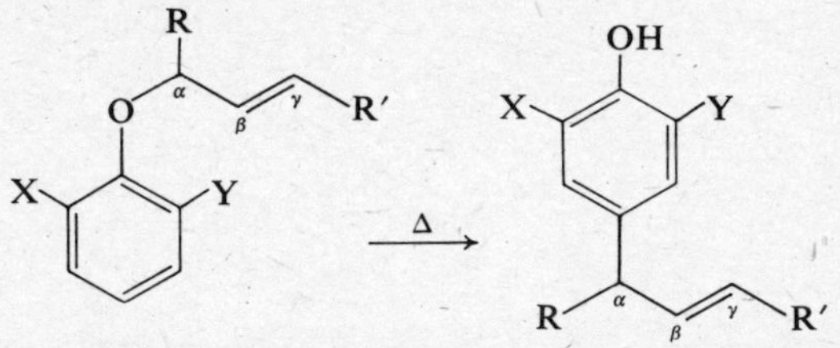

(83) (84)

ring in the formation of the o-allylphenol. This has been shown both by study of the rearrangement of allyl ethers substituted on the α- or γ-carbon atoms (i.e., 79 → 80) and, more elegantly, by use of isotopically labelled compounds (e.g., 81 → 82). In the *para*-rearrangement, on the other hand, it is the α-carbon atom of the allyloxy group that becomes bonded to the benzene ring in the *p*-allylphenol product (i.e., 83 → 84).

The concerted intramolecular nature of the process has been established, for example, through study of the rearrangement of optically active substrates. Thus, an optically active form of the ether (79; R = R′ = Me; note that C_α is asymmetric) gave [28b] an optically active form of the phenol (80; R = R′ = Me; note that C_γ is asymmetric). In this reaction, the chiral centre is transferred from one carbon atom in the reactant to another in the product.

The mechanisms now generally accepted for the Claisen rearrangements are outlined in Scheme 9. In the *ortho*-rearrangement, migration of the allyl group occurs by a [3,3] sigmatropic shift of the C_α—O bond to a C_γ—C_{ortho} location with a concomitant shift of the allylic

Scheme 9. Claisen Rearrangements

ortho-Rearrangement

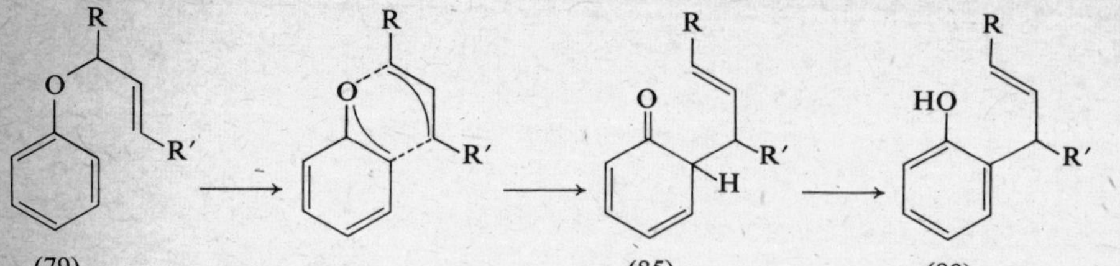

R O R' (79) → R O R' → R O R' H (85) → R HO R' (80)

para-Rearrangement

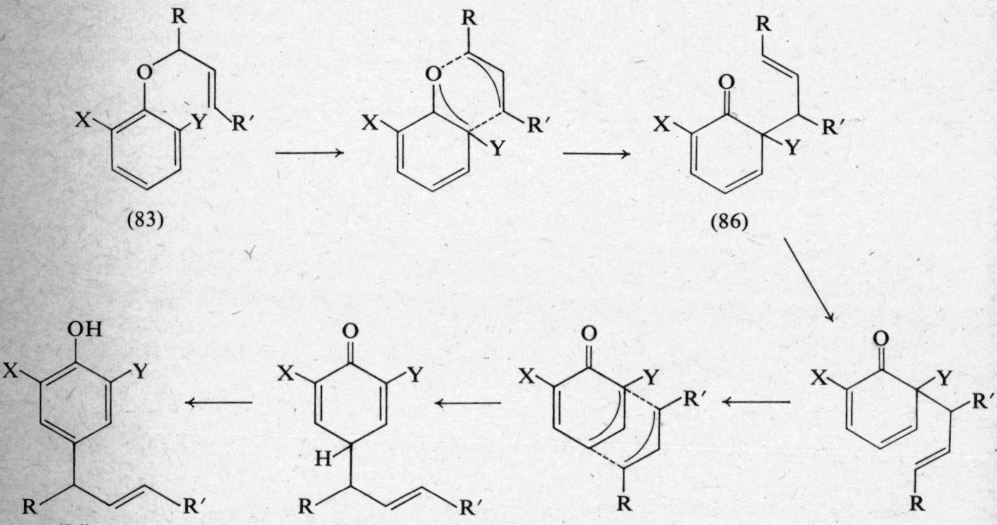

R O X Y R' (83) → R O X R' Y → R O X R' Y (86) → O X Y R' R (86) → O X Y R' R → O X Y H R R' (87) → OH X Y R R' (84)

double bond. Tautomeric shift of the labile *ortho*-hydrogen atom in the *o*-dienone intermediate (85) thereby formed occurs readily, giving the *o*-allylphenol product (80). Such keto → enol tautomerism obviously cannot occur in the rearrangement of substrates lacking an *ortho*-hydrogen atom. However, the *o*-dienone intermediate (86), produced in these cases, may undergo a Cope rearrangement (*vide supra*) involving [3,3] sigmatropic shift of the C_{γ}—C_{ortho} bond to a C_{α}—C_{para} location (see Scheme 9). The *p*-dienone intermediate (87) formed by this process may then tautomerise to the observed *p*-allylphenol product (84).

The mechanistic pathways outlined in Scheme 9 account nicely for the observed overall course of the Claisen *ortho*- and *para*-rearrangements. Direct evidence for the intervention of *o*-dienone intermediates has been obtained. For example, it has been shown [29] that when the thermal rearrangement of the allyl xylenyl ether (88) is carried out in the presence of maleic anhydride, which is a powerful dienophile, a small amount of the Diels–Alder adduct (89) is formed in addition to the *para*-rearrangement product (90).

An *o*-dienone (92) has, in fact, been synthesised by the reaction of the phenoxide (91) with allyl bromide. At relatively low temperatures (75–100°), this dienone rearranges, by a *para*-rearrangement, to a mixture of phenol (93) and the ether (94). Since the *para*-rearrangement of this ether occurs at more elevated temperature (i.e., 94 → 93), this result shows that the first stage in the Claisen rearrangement (i.e., production of an *o*-dienone) must be a reversible process [30].

The reversibility of *o*-dienone formation in the first step of the Claisen rearrangement (i.e., 79 ⇄ 85; see Scheme 9) has also been established by the results found [31] for the thermolyses of the allyl aryl ethers (95 and 96) bearing allylic substituents at the *ortho*-positions of the aryl ring (see Scheme 10). Each ether yields a mixture of phenols (99 and 100) on heating. The *o*-dienone intermediate (97), formed from the symmetrically substituted ether (95), must therefore be capable of reversible rearrangement to the isomeric ether (96) and

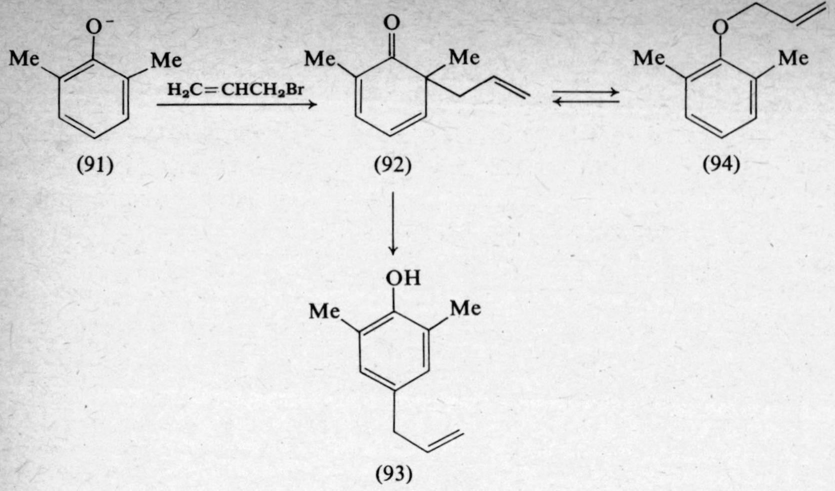

$H_2C{=}CHCH_2Br$

Me Me O⁻ (91)

Me Me O (92)

Me Me O (94)

OH Me Me (93)

Scheme 10

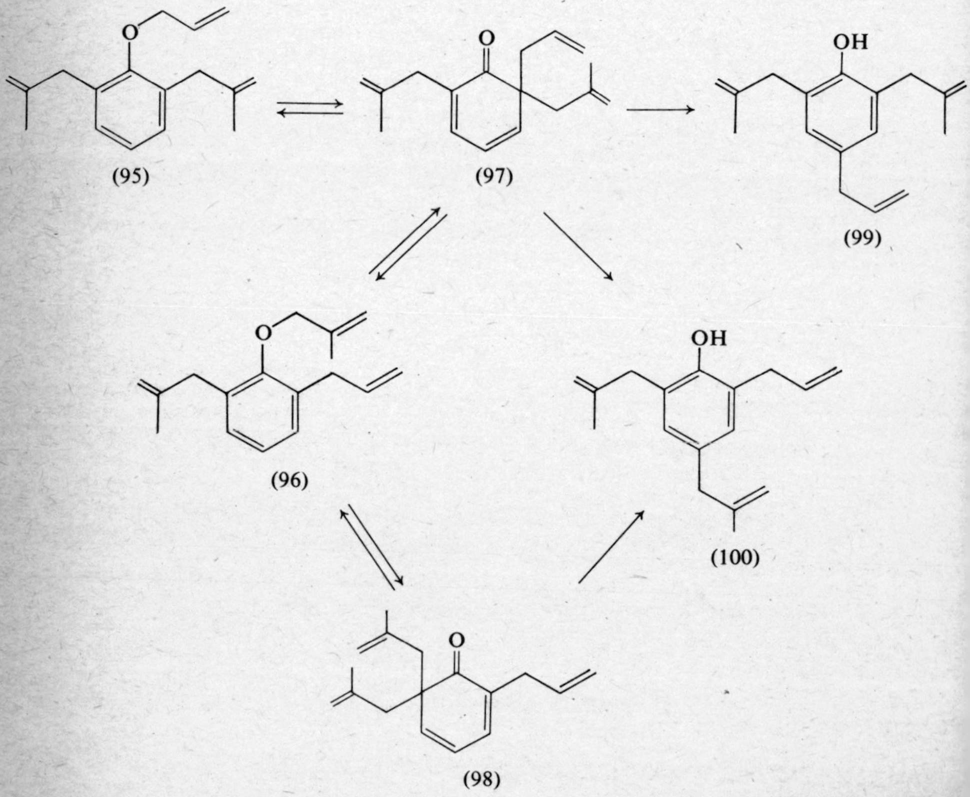

(95)

O (97)

OH (99)

O (96)

OH (100)

O (98)

thence to the phenol (100) via the *o*-dienone intermediate (98). In other words, each precursor (95 or 96) must be capable of furnishing both *o*-dienones (97 and 98), and this can occur only if the ether to *o*-dienone reaction is reversible.

As in the case of the Cope rearrangement (*vide supra*), two transition-state geometries (cf. Scheme 8) are possible in principle for the first step in the Claisen rearrangement. Recent work [28c] has established a similar preference for the chair-like rather than the boat-like arrangement.

Examples of out-of-ring migration of the allyl group in suitably substituted allyl aryl ethers have been observed (see Scheme 11). For example, Claisen discovered [32] that ethers of the type (101) undergo

Scheme 11. Out-of-ring Claisen Rearrangements

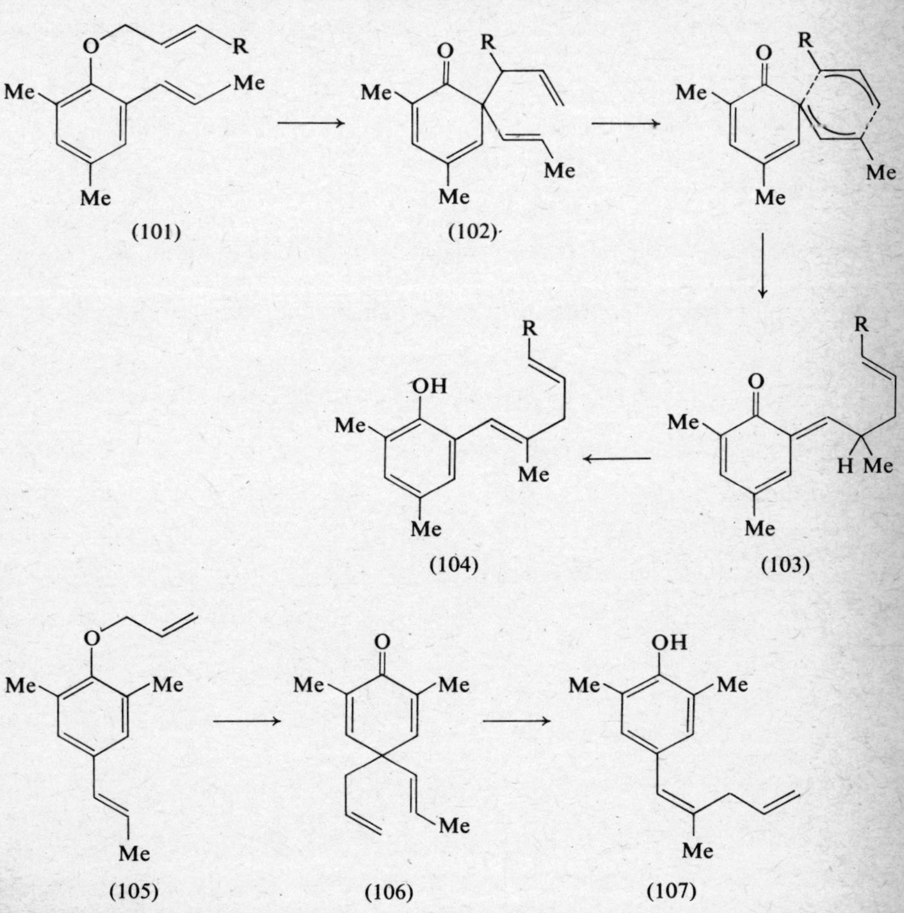

(101) (102) (103) (104)

(105) (106) (107)

thermal rearrangement to phenols (104) by migration of the allyl group from oxygen to the β-carbon atom of the *o*-propenyl substituent. The presence of *ortho*- and *para*-substituents in these ethers precludes normal Claisen rearrangement which, as we have seen, requires the production of an enolisable *o*- or *p*-dienone intermediate. The mechanism of this rearrangement has been established [33]. The *o*-dienone intermediate (102), formed by normal [3,3] sigmatropic rearrangement of the ether (101), may undergo a Cope rearrangement to give an *o*-quinonoid structure (103) which readily enolises to form the observed product (104). A related out-of-ring migration of an allyl group to the β-carbon atom of a *p*-propenyl substituent has also been demonstrated [34] (i.e., 105 → 107). In this case, the *p*-dienone intermediate (106), formed from the ether (105) by two successive [3,3] sigmatropic shifts (cf. mechanism of the Claisen *para*-rearrangement), may undergo a further Cope rearrangement leading to the *p*-quinonoid tautomer of the product (107).

In the Claisen *ortho*-rearrangements discussed so far, it has been shown that it is the γ-carbon atom of the migrating allyl group which becomes attached to the benzene ring (see Scheme 9). However, abnormal rearrangements have also been found in which a different behaviour is observed. For example, pent-2-enyl phenyl ether (108) rearranges on heating to give *o*-(pent-2-en-4-yl)phenol (110), a reaction first reported [35] in 1936. More recent research has established [36] that in fact the normal *ortho*-rearrangement product (109)

Scheme 12. Abnormal Claisen Rearrangement

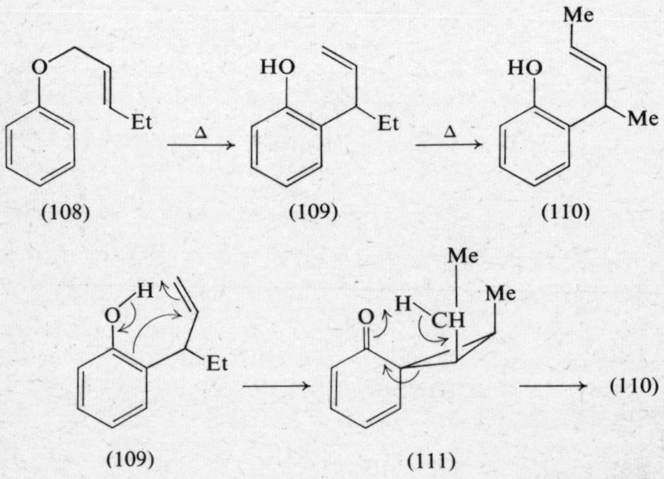

is formed in the reaction, and that the abnormal product (110) is derived by thermal rearrangement of this phenol. Isotopic labelling experiments [37] have been interpreted in favour of a mechanism in which the phenol (109) is converted to a spiro-cyclopropane intermediate (111) as shown (Scheme 12). This reaction occurs by transfer of the phenolic hydrogen atom to the terminal vinylic carbon atom of the pentenyl group, with concomitant formation of a three-membered ring. This ring may then open by a concerted process involving transfer to oxygen of a methylene hydrogen atom of the ethyl group, and regeneration of the benzenoid ring.

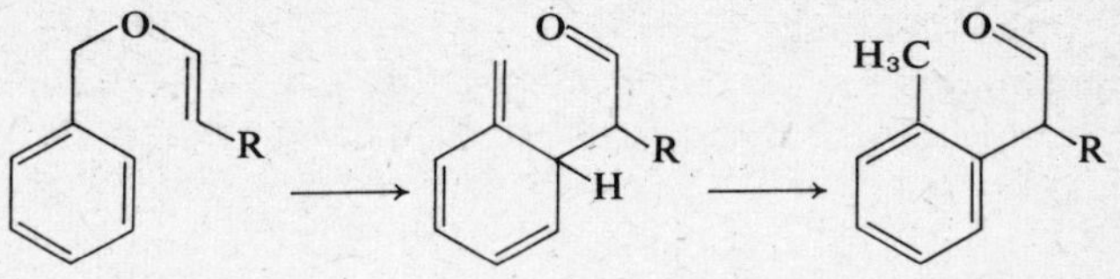

(112) (113)

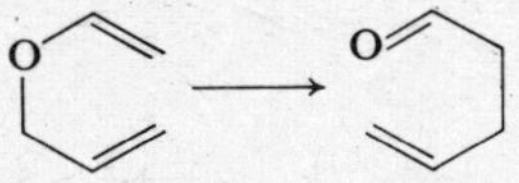

(114) (115) 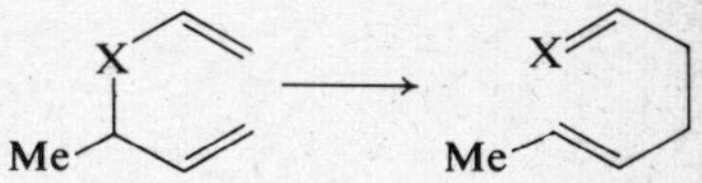 (116)

The Claisen *ortho*-rearrangement (73 → 74) is a special example of a thermal rearrangement which is general for substrates which incorporate an allyl vinyl ether fragment in their molecular structure. In the *ortho*-rearrangement, a benzene ring serves as the vinylic group in the sigmatropic process (i.e., 79 → 85). Related rearrangements are known in which a benzyl system serves as the allylic group (e.g., 112 → 113). The minimum structure requirement for the reaction is shown by the rearrangement of allyl vinyl ether (114) which is itself converted to pent-4-enal (115) at 150°. The close structural relationship between this type of reaction and the Cope rearrangement (31 → 32) is readily apparent.

For structurally similar systems, the free energy of activation for the Claisen rearrangement is usually significantly lower than that for the corresponding Cope rearrangement, such that the former transformation occurs at a much faster rate at a given temperature [38]. For example, the ether (116; X = O) rearranges 10,000 times faster than the hydrocarbon (116; X = CH_2) at 150°. From consideration of the mechanisms of these rearrangements, it is understandable that this difference in free energy of activation derives from the enthalpy rather than the entropy component. The change in the degree of ordering of the system on passing from reactant to transition state is similar for both processes.

References

1. R. B. Woodward and R. Hoffmann, *J. Amer. Chem. Soc.*, 1965, **87**, 2511.
2. R. W. Murray and M. L. Kaplan, *J. Amer. Chem. Soc.*, 1966, **88**, 3527.
3. W. R. Roth, *Annalen*, 1964, **671**, 25.
4. W. R. Roth, J. König, and K. Stein, *Chem. Ber.*, 1970, **103**, 426.
5. J. A. Berson and G. L. Nelson, *J. Amer. Chem. Soc.*, 1967, **89**, 5503.
6. (a) W. von E. Doering and H. D. Roth, *Tetrahedron*, 1970, **26**, 2825. (b) M. J. S. Dewar and J. S. Wasson, *J. Amer. Chem. Soc.*, 1971, **93**, 3081; J. J. Gajewski, *ibid.*, 1971, **93**, 4450.
7. T. M. Brennan and R. K. Hill, *J. Amer. Chem. Soc.*, 1968, **90**, 5614.
8. (a) A. C. Cope and E. M. Hardy, *J. Amer. Chem. Soc.*, 1940, **62**, 441. (b) A. C. Cope *et al.*, *J. Amer. Chem. Soc.*, 1941, **63**, 1843, 1852; 1943, **65**, 1999; 1944, **66**, 1684; 1947, **69**, 1893; 1956, **78**, 2547, 2551.
9. C. A. Grob and P. Schiess, *Angew. Chem.*, 1958, **70**, 502.
10. E. Vogel, *Annalen*, 1958, **615**, 1; *Angew. Chem.*, 1960, **72**, 21.
11. E. Vogel, K. H. Ott, and K. Gajek, *Annalen*, 1961, **644**, 172.
12. C. Cupas, W. E. Watts, and P. von R. Schleyer, *Tetrahedron Letters*, 1964, 2503; J. M. Brown, *Chem. Comm.*, 1965, 226.
13. G. W. Klumpp, J. W. F. K. Barnick, A. H. Veefkind and F. Bickelhaupt, *Rec. Trav. Chim.*, 1969, **88**, 766.
14. M. Rey and A. S. Dreiding, *Helv. Chim. Acta*, 1965, **48**, 1985.
15. S. J. Rhoads and R. D. Cockroft, *J. Amer. Chem. Soc.*, 1969, **91**, 2815.
16. S. F. Reed, *J. Org. Chem.*, 1965, **30**, 1663.
17. E. Vogel, R. Erb, G. Lenz, and A. A. Bothner-By, *Annalen*, 1965, **682**, 1.
18. H. A. Staab *et al.*, *Chem. Ber.*, 1965, **98**, 2681, 2691, 2701, 3479.
19. W. von E. Doering and W. R. Roth, *Tetrahedron*, 1963, **19**, 715.
20. A. C. Cope, A. C. Haven, F. L. Ramp, and E. R. Trumbull, *J. Amer. Chem. Soc.*, 1952, **74**, 4867.
21. W. von E. Doering, B. M. Ferrier, E. T. Fossel, J. H. Hartenstein, M. Jones, G. W. Klumpp, R. M. Rubin, and M. Saunders, *Tetrahedron*, 1967, **23**, 3943.
22. J. B. Lambert, *Tetrahedron Letters*, 1963, 1901.
23. G. Schröder, *Chem. Ber.*, 1964, **97**, 3140, 3150.
24. W. von E. Doering and W. Roth, *Tetrahedron*, 1962, **18**, 67.
25. R. B. Woodward and T. J. Katz, *Tetrahedron*, 1959, **5**, 70.
26. L. Claisen, *Ber.*, 1912, **45**, 3157.

27. C. F. H. Allen and J. W. Gates, *Org. Synth. Coll. Vol.* **3**, 1955, 418.
28. (a) J. F. Kincaid and D. S. Tarbell, *J. Amer. Chem. Soc.*, 1939, **61**, 3085; H. Schmid and K. Schmid, *Helv. Chim. Acta*, 1952, **35**, 1879; 1953, **36**, 489; S. J. Rhoads, R. Raulins, and R. D. Reynolds, *J. Amer. Chem. Soc.*, 1953, **75**, 2531; 1954, **76**, 3456.
(b) E. R. Alexander and R. W. Kluiber, *J. Amer. Chem. Soc.*, 1951, **73**, 4304.
(c) G. Frater, A. Habich, H.-J. Hansen, and H. Schmid, *Helv. Chim. Acta*, 1969, **52**, 335.
29. H. Conroy and R. A. Firestone, *J. Amer. Chem. Soc.*, 1953, **75**, 2530; 1956, **78**, 2290.
30. D. Y. Curtin and R. J. Crawford, *J. Amer. Chem. Soc.*, 1957, **79**, 3156.
31. D. Y. Curtin and H. W. Johnson, *J. Amer. Chem. Soc.*, 1956, **78**, 2611.
32. L. Claisen and E. Tietze, *Annalen*, 1926, **449**, 89.
33. W. M. Lauer and D. W. Wujciak, *J. Amer. Chem. Soc.*, 1956, **78**, 5601; K. Schmid, P. Fahrni, and H. Schmid, *Helv. Chim. Acta*, 1956, **39**, 708.
34. A. Nickon and B. R. Aaronoff, *J. Org. Chem.*, 1964, **29**, 3014.
35. W. M. Lauer and W. F. Filbert, *J. Amer. Chem. Soc.*, 1936, **58**, 1388.
36. E. N. Marvell, D. R. Anderson, and J. Ong, *J. Org. Chem.*, 1962, **27**, 1109.
37. A. Habich, R. Barner, W. von Philipsborn, R. M. Roberts, and H. Schmid, *Helv. Chim. Acta*, 1962, **45**, 1943; 1965, **48**, 1297.
38. H. M. Frey and R. Walsh, *Chem. Rev.*, 1969, **69**, 103.

Review Articles

The review articles listed at the end of Chapter 7 deal with sigmatropic rearrangements. More specialised reviews are given below.

Cope and Claisen rearrangements

S. J. Rhoads in *Molecular Rearrangements*, Part 1, Chapter 11, ed. de Mayo, Interscience (New York), 1963.

Claisen rearrangement

D. S. Tarbell, *Org. Reactions*, 1944, **2**, 1.

A. Jefferson and F. Scheinmann, *Quart. Rev.*, 1968, **22**, 391.

H.-J. Hansen and H. Schmid, *Chem. Brit.*, 1969, **5**, 111.

Valence tautomerism

G. Schröder, J. F. M. Oth, and R. Merényi, *Angew. Chem. Internat. Edn.*, 1965, **4**, 752.

Sigmatropic rearrangements

J. A. Berson, *Accounts of Chem. Research*, 1968, **1**, 152.

9

Some Aromatic Rearrangements

The transformations discussed in this chapter are mostly of the type $C_6H_5NHX \rightarrow XC_6H_4NH_2$, in which an atom or group migrates from nitrogen to the *o*- or *p*-position of an attached benzene ring. They take place under acid conditions, of widely varied degrees of severity; the migrating entity (X) may be a halogen atom, a hydroxyl group, a nitroso- or nitro-group or other nitrogenous residue, or an alkyl group.

In spite of their superficial similarity, the reactions are not all of a kind. In some cases the 'obvious' two-stage mechanism seems to hold, i.e., separation of X^+ or a related electrophilic product, followed by substitution in the aromatic ring. In others, such a sequence is not followed, and each case must be discussed separately. In this series we encounter the unusual possibility of an *inter*molecular process which does not involve liberation of the migrating group: if a trace of $PhNH_2$ is present or produced in the reaction, and if $Ph\overset{+}{N}H_2X$ is an effective electrophilic substituting agent, the following sequence can occur:

$$Ph\overset{+}{N}H_2X + C_6H_5{-}NH_2 \longrightarrow$$

$$PhNH_2 + (X)(H)C_6H_4{=}\overset{+}{N}H_2 \longrightarrow X{-}C_6H_4{-}NH_2$$

the regenerated aniline continues a chain reaction.

Dewar [1], advocate of the π-complex formulation of electrophilic substitution, applied similar ideas to these rearrangements. Heterolytic detachment of X^+ from $Ph\overset{+}{N}H_2X$ is immediately followed by π-complexing (1). X^+ is attached *peripherally* to the nucleus, and while its passage round the ring is easy there is a real, if low, energy barrier

$$H_2N\text{-}C_6H_5 + X^+ \longrightarrow H_2N{\rightarrow}X^+ \longrightarrow H_2N\text{=}(\ldots)X^+ \text{ etc.}$$

(1)

between each step. If, then, there is a strong tendency for the π-complex to yield the stable, normally covalent product, X^+ may not often get beyond the *ortho*-position. This concept has been applied most notably to the nitramine rearrangement and to the benzidine change (q.v.).

N-Haloanilides (the Orton Rearrangement [2])

Acetanilide, with hypochlorite, gives the *N*-chloro derivative PhNClAc, which is isomerised smoothly to *p*- and some *o*-chloroacetanilide by hydrogen chloride in aqueous acetic acid. The following sequence is well established:

$$PhNClAc + H^+ \longrightarrow Ph\overset{+}{N}HClAc$$

$$Ph\overset{+}{N}HClAc + Cl^- \longrightarrow PhNHAc + Cl_2$$

$$Cl_2 + PhNHAc \longrightarrow ClC_6H_4NHAc + HCl$$

The above sequence is supported by the following experimental observations.

(a) Gaseous chlorine can be aspirated from the solution and acetanilide isolated.

(b) The ratio of *o*- to *p*-chloroacetanilide produced is the same for the isomerisation as for the direct chlorination of acetanilide under the same conditions.

(c) In the same way, *N*-bromoacetanilide and hydrogen bromide give *o*- and *p*-bromoacetanilide, also formed from *N*-chloroacetanilide/HBr and from *N*-bromoacetanilide/HCl. These last two cases presumably involve the formation of BrCl, known to be a *brominating* agent. Other strong acids, e.g. sulphuric acid, are not effective under these conditions.

(d) When, for example, 2,4,*N*-trichloroacetanilide is isomerised to 2,4,6-trichloroacetanilide, the addition of a substance with a more reactive benzene ring (acetanilide or anisole) can trap the liberated chlorine.

(e) Studies of the kinetics of the rearrangement of a number of examples, in aqueous acetic acid or in water, gave results fully accounted for by the reversible *N*-chlorination and irreversible *C*-chlorination of acetanilide, without any 'direct' rearrangement.

It is possible that other mechanisms may operate under other conditions. Bell observed general acid catalysis, particularly by carboxylic acids, for isomerisation in chlorobenzene, with no detectable liberation of halogen, and inferred a one-stage intramolecular process. Soper pointed out [3] the very likely production of acyl hypochlorite, RCO(OCl), a good chlorinating agent, and the matter remains *sub judice*. The rearrangement can also be effected photochemically, probably by a homolytic mechanism.

Rearrangement of Diazoamino-compounds

This process appears to be comparable with the Orton rearrangement. Under the influence of acids, most effectively by arylamine hydrochloride in arylamine as solvent, the diazoamino-compound is isomerised to an aminoazo-compound:

$$\mathrm{C_6H_5{-}NHN{=}NPh} \longrightarrow \mathrm{PhN{=}N{-}C_6H_4{-}NH_2}$$

$$\mathrm{C_6H_5NHN{=}NPh} \underset{}{\overset{H^+}{\rightleftharpoons}} \mathrm{Ph\overset{+}{N}H_2N{=}NPh} \rightleftharpoons \mathrm{PhNH_2} + \mathrm{\overset{+}{N}_2Ph} \xrightarrow{-H^+} \mathrm{PhN{=}N{-}C_6H_4{-}NH_2}$$

The above mechanism was, in essentials, proposed by Friswell and Green in 1885. The fission of a diazoamino-compound into a diazonium ion and arylamine hydrochloride by excess of acid is well established, and exploited technically in the use of these compounds as 'stabilised diazonium ions'. In the rearrangement, the diazonium ion can attack aniline reversibly at the nitrogen atom, and irreversibly at benzenoid carbon. The intermediacy of a diazonium ion is supported by the facts that the rearrangement of *p*-diazoaminotoluene in aniline/aniline hydrochloride gives the 'mixed' aminoazo-compound $MeC_6H_4N{=}NC_6H_4NH_2$, and that diazoaminobenzene in presence of dimethylaniline yields dimethylaminoazobenzene. Following Goldschmidt [4] it is suggested, alternatively, that the coupling agent is protonated diazoaminobenzene (2).

$$\underset{(2)}{\mathrm{Ph\overset{+}{N}H_2{-}N{=}N{-}Ph}} + \mathrm{C_6H_5{-}NH_2} \xrightarrow{-H^+} \mathrm{PhNH_2} + \mathrm{PhN{=}N{-}C_6H_4{-}NH_2}$$

The Hofmann–Martius Rearrangement)

In this rearrangement, recorded in 1872, an alkyl group attached to ammonium nitrogen migrates into the *o*- and *p*-positions of the aromatic ring. Thus, at 250–300°, methylaniline hydrobromide yields *o*- and *p*-toluidine hydrobromides. That the reaction is not purely intramolecular is shown by the behaviour of, for example, *n*-butylaniline hydrobromide (3), which affords *N*,*p*-dibutylaniline and aniline as well as *p*-butylaniline. Hickinbottom [5] reasonably inferred the separation of alkyl cation or its equivalent, followed by nuclear substitution.

$$C_6H_5\overset{+}{N}H_2Bu\,Br^- \text{ (3)} \longrightarrow p\text{-}BuC_6H_4NH_2 + p\text{-}BuC_6H_4NHBu + C_6H_5NH_2$$

Hickinbottom has recorded further complications. By-products of the reaction include alkyl halide and alkene. When the initial secondary amine carries an alkyl group which would rearrange as cation (e.g., isoamyl), the main product is *p*-tert. amylaniline and the by-product alkene is (also rearranged) trimethylethylene; but the halide is unrearranged isoamyl bromide. It is suggested that the initial process parallels that in the Orton rearrangement:

$$C_6H_5\overset{+}{N}H_2R + Hal^- \longrightarrow C_6H_5NH_2 + RHal$$

Then, in the strongly polar medium constituted by the fused salt, the halide ionises; R^+ can rearrange to the more stable tertiary carbonium ion, and this can either lose a proton to give the alkene, or attack the aromatic nucleus. Hickinbottom has discovered a further, unexplained, complexity. If an *N*-alkylaniline is heated with a metallic halide such as $CoBr_2$, it is rearranged to *p*-alkylaniline without reorganisation of the migrating group; *N*-isoamylaniline gives *p*-isoamylaniline.

Rearrangement of Nitrosamines

This was first recorded by Otto Fischer and Hepp in 1886. When an alkylarylnitrosamine is treated with hydrogen halide in alcohol or ether, the nitroso group is transferred to the *p*- or occasionally to an *o*-position of the benzene ring. Circumstantial evidence led to the general acceptance of the view [6] that nitrosyl halide is split off and that this nitrosates the nucleus.

$$\text{PhNMeNO} + \text{HHal} \longrightarrow \text{PhNHMe} + \text{NOHal}$$

Thus, when the reaction is conducted in presence of urea as a trap for nitrosyl chloride, the product is the denitrosated secondary amine; in presence of dimethylaniline, *p*-nitrosodimethylaniline is formed. The hydrogen halides are more effective than other strong acids.

This neatly clean-cut conclusion was disturbed when it was found [7] that, if methylphenylnitrosamine is rearranged in the presence of ^{15}N-labelled nitrite, little or no ^{15}N enters the *p*-nitroso-product. It is unlikely that any $\overset{+}{N}O$ or NOCl, liberated from the nitrosamine and kinetically free, would fail to equilibrate promptly with labelled nitrite. It would seem that the reaction does not proceed, in the main, according to the formerly accepted mechanism, but is either a truly intramolecular process or involves nitrosation of one molecule by another (compare Goldschmidt's mechanism for the rearrangement of diazoamino-compounds).

Rearrangement of Nitramines

This case (Bamberger, 1893) is at least in part fundamentally different from those so far considered, and its mechanism is still the subject of controversy. When phenylnitramine, $PhNHNO_2$, is treated with fairly strong sulphuric acid, the greatly predominating product is *o*-nitroaniline. Only in exceptional cases can a nitro group be transferred from a nitramine to a second aromatic ring. Hughes and Jones, following earlier work by Holleman, compared the products obtained by the action of 85 per cent sulphuric acid on (a) phenylnitramine and (b) aniline nitrate (see Table 1). Clearly the nitramine rearrangement

Table 1. Production of Nitroanilines

Substrate	% *ortho*	% *meta*	% *para*
Phenylnitramine	93	0	7
Aniline nitrate	6	34	59

does not proceed wholly by formation of kinetically free HNO_3 or $\overset{+}{N}O_2$. It was further shown that, when the rearrangement is conducted in presence of $^{15}NO_3H$, no ^{15}N is taken up [8]. White, however, observed [9] transfer of ^{15}N when $PhNMeNO_2$ was rearranged in presence of $FC_6H_4NMe^{15}NO_2$.

Dewar urges that the clearly intramolecular character of the change and the very predominant *ortho*-migration in phenylnitramine give strong support to a π-complex mechanism.

A well-known case of easy, strictly intramolecular, *ortho*-migration from side-chain to nucleus is the Claisen rearrangement of allyl aryl ethers (see Chapter 8). On this consideration, Hughes [8] proposed an imaginative interpretation of the nitramine change (Scheme 1).

Scheme 1

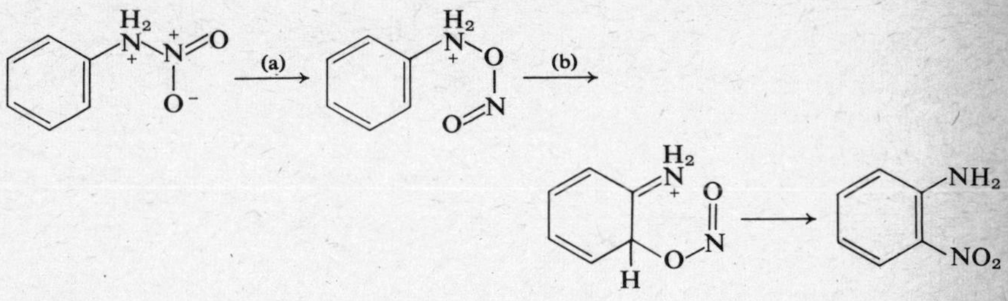

(a) (b)

Stage (a) is an electrophilic rearrangement incited by the electrostatic repulsion of the two cationic nitrogen atoms. Step (b) is the exact counterpart of the *ortho*-Claisen shift. The final step involves, with simultaneous or subsequent loss of a proton, a second electrophilic migration in which the migrating entity will be the rather stable radical-ion (4) (cf. Chapter 5).

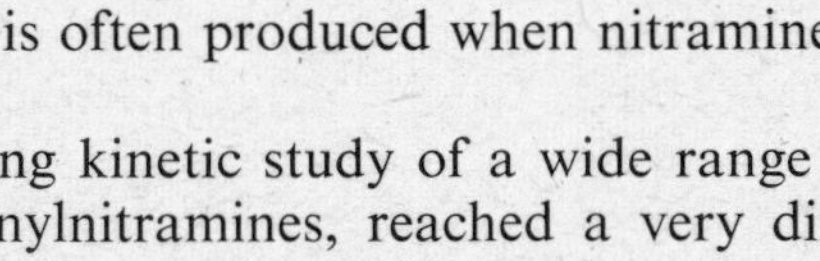

(4)

Consistent with the Hughes theory is the observation that *nitrous* acid or its equivalent is often produced when nitramines are treated with acid.

White, following kinetic study of a wide range of nuclear-substituted methylphenylnitramines, reached a very different conclusion [10]. The reaction is strongly accelerated by electron-donating substituents in the ring (about 900-fold by *p*-OMe; about 400-fold retardation by *p*-NO_2). There is good agreement with a Hammett σ^+ formula

[11] with the high negative ρ-value, $-3{\cdot}7$. These results are regarded as indicating that in the transition state a high degree of 'onium character is developed on the atom adjacent to the benzene ring, and that the process is strongly promoted conjugatively by a suitable electron-donating substituent. This would negate the Dewar and the Hughes proposals and be consonant with a homolytic mechanism like that of Scheme 2. Although less outstanding, in general, than in the simplest case of phenylnitramine, the predominant *ortho*-migration is accommodated in the same way as in Dewar's π-complex theory (p. 192).

Scheme 2

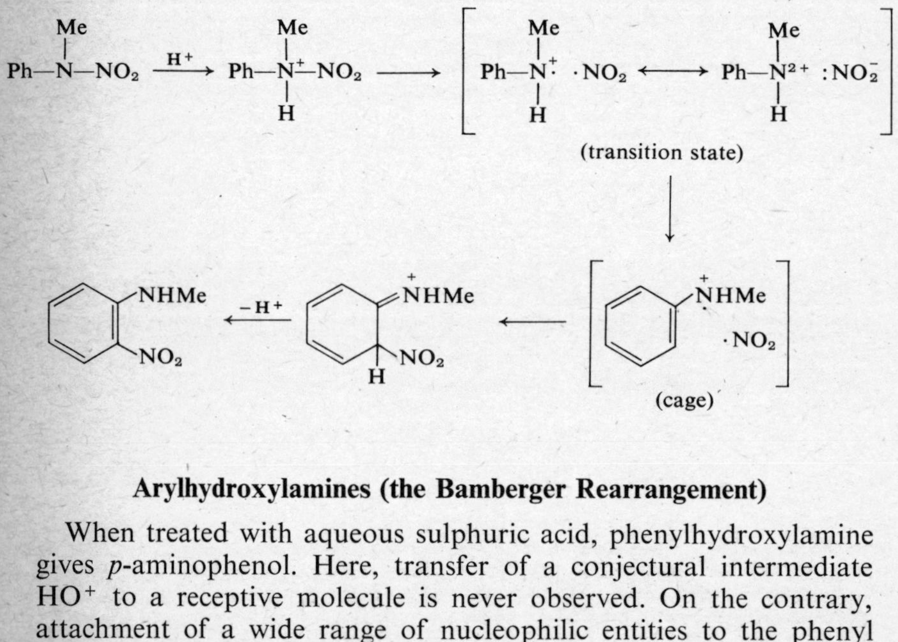

$$\mathrm{Ph{-}N(Me){-}NO_2} \xrightarrow{H^+} \mathrm{Ph{-}\overset{+}{N}(Me)(H){-}NO_2} \longrightarrow \left[\mathrm{Ph{-}\overset{+}{N}{\cdot}(Me)(H)\ \cdot NO_2} \longleftrightarrow \mathrm{Ph{-}N^{2+}(Me)(H)\ :NO_2^-}\right]$$

(transition state)

$$\left[\mathrm{C_6H_5\overset{+}{N}HMe\ \ \cdot NO_2}\right] \text{(cage)} \longrightarrow \text{(cyclohexadienyl } {=}\overset{+}{N}HMe,\ H,\ NO_2) \xrightarrow{-H^+} o\text{-}\mathrm{NO_2C_6H_4NHMe}$$

Arylhydroxylamines (the Bamberger Rearrangement)

When treated with aqueous sulphuric acid, phenylhydroxylamine gives *p*-aminophenol. Here, transfer of a conjectural intermediate HO^+ to a receptive molecule is never observed. On the contrary, attachment of a wide range of nucleophilic entities to the phenyl nucleus is realised. In methanolic or ethanolic acid the product is *p*- (and *o*-) anisidine or phenetidine; with phenol as an addendum, its *p*-carbon atom becomes attached to the *p*-carbon of the phenylhydroxylamine, giving $HOC_6H_4C_6H_4NH_2$. Rearrangement of phenylhydroxylamine in the presence of hydrochloric acid gives much *p*-chloroaniline.

It is suggested (Bamberger, Yukawa) that the significant intermediate is the mesomeric cation (5) resulting from loss of water by the *O*-protonated hydroxylamine. The intermediate cation is the same as that postulated for the acid-induced rearrangement of aryl azides (Chapter 3) and the products of both reactions are strikingly similar.

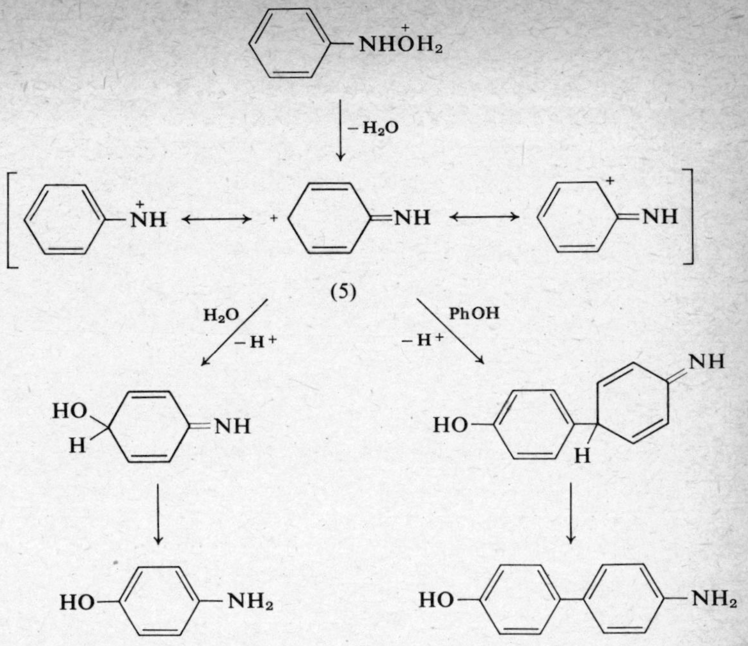

Wallach rearrangement. The rearrangement of azoxybenzenes in the presence of strong sulphuric acid, recorded by Wallach in 1880 and recently reviewed by Buncel, seems superficially akin to the transformation of arylhydroxylamines:

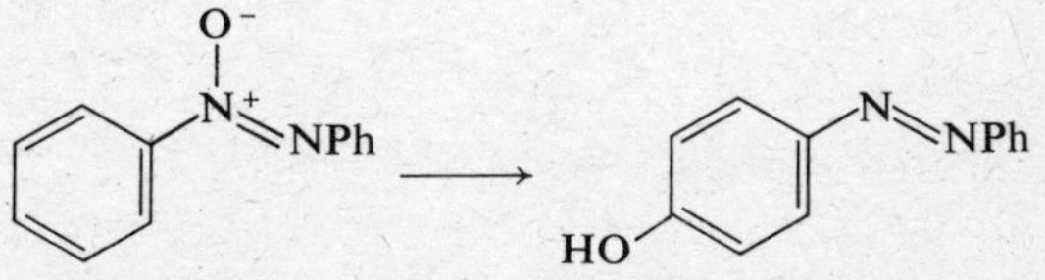

The mechanism of the change is still controversial. A corresponding *ortho*-migration can be effected photochemically.

The Benzidine Rearrangement

This reaction was first recorded by Hofmann as early as 1863, and studied in great detail [12] by Jacobson in the thirty years up to 1922. When hydrazobenzene (6) is treated with acid reagents, e.g., aqueous ethanolic hydrogen chloride, it is rapidly converted into a mixture of benzidine (7) (70–75%) and diphenyline (8) (25–30%).

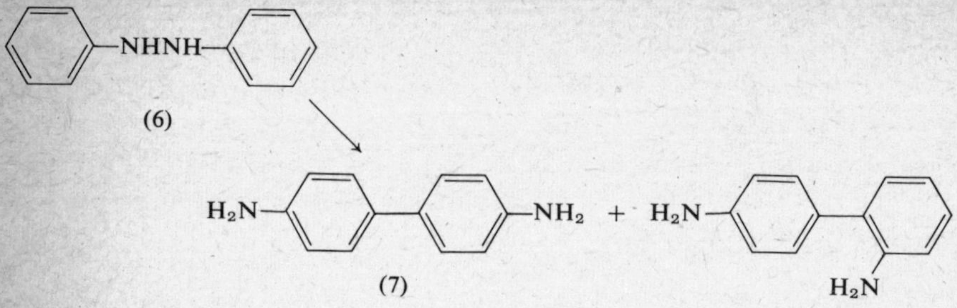

The study of a wide range of substituted hydrazobenzenes has disclosed in various cases the formation of products of the types shown in Table 2.

Table 2

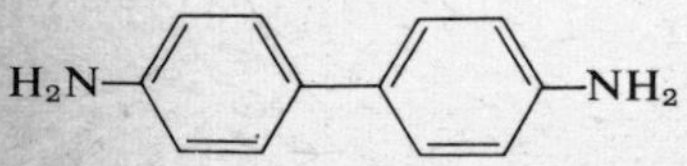

Benzidine 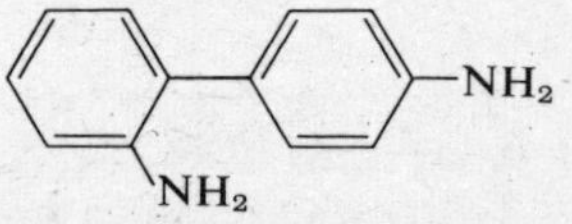   Diphenyline 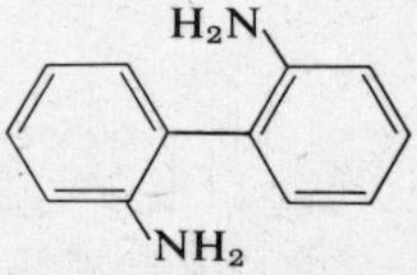   *ortho*-Benzidine

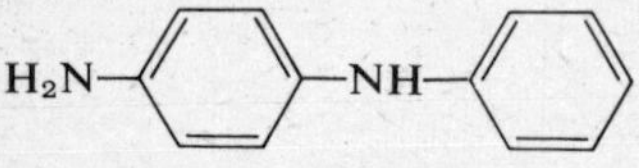

para-Semidine 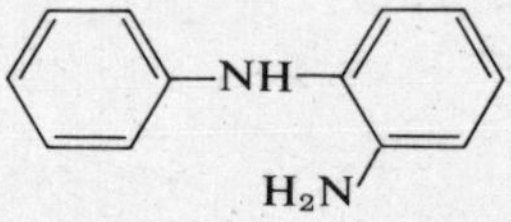   *ortho*-Semidine

(Aniline and azobenzene may also be formed by disproportionation:

$$2PhNHNHPh \longrightarrow 2PhNH_2 + PhN{=}NPh)$$

The benzidine is often the only major product; its formation may involve expulsion of a 4-substituent (CO_2H, SO_3H, sometimes Cl or OAc) in the parent hydrazobenzene. Substituted hydrazobenzenes seldom yield more than traces of *ortho*-benzidine, whereas hydrazonaphthalenes give much *ortho*-benzidine but no diphenyline. A hydrazobenzene with one firmly fixed 4-substituent cannot give a benzidine, and may afford either a diphenyline or an *ortho*-semidine; with two such substituents the product is the *ortho*-semidine. A *para*-semidine is less often a major product; it has been asserted, and strongly denied, that its formation requires the presence of a heavy-metal ion and should be regarded as an independent phenomenon.

Dewar has formulated [1] rules summarising the products observed in the rearrangement of hydrazobenzenes. Protonation takes place on the more basic nitrogen atom, and the attached ring is designated A.

(a) The preferred order of products is benzidine > diphenyline ⩾ semidine.

(b) A diphenyline is the main product only if ring A has a free *p*-position; it is formed by linkage to an *o*-position of ring B.

(c) If a semidine is formed, ring A carries the free amino group. A single electron-releasing *p*-substituent (ring A) leads to formation of an *o*-semidine, whereas an electron-attracting group (ring B) leads to a diphenyline.

It is tempting to formulate the benzidine changes as twofold side-chain-to-nucleus migrations (i.e. $PhNHN'HPh \rightarrow H_2N'C_6H_4NHPh \rightarrow H_2NC_6H_4C_6H_4N'H_2$); semidines, however, are unaffected by the usual conditions of the benzidine change and so cannot be intermediates.

Early workers also envisaged fission of the N—N bond (essentially homolytic, Tichwinsky; heterolytic, Stieglitz), giving fragments at least tacitly regarded as kinetically free, that combine to give the observed products. Such schemes were refuted (Ingold, Wheland) by mixture experiments, of which the most conclusive was the simultaneous rearrangement of 2,2′-dimethylhydrazobenzene and 2-^{14}C-methylhydrazobenzene; the dimethylbenzidine produced was not radioactive [13].

Jacobson recorded, semi-quantitatively, the products obtained from some eighty well-diversified substituted hydrazobenzenes, and this work has since been extended and made more precise. The more recent, very elaborate kinetic studies cannot be discussed here in adequate detail; Ingold has provided [14] a very full review. The ratio of products is not greatly affected by changes in solvent composition, and still less by varying concentration of acid. Earlier kinetic study of several cases gave rates of second-order in hydrogen-ion concentration. Later, quite a few examples were found of one-proton, and some of mixed kinetics.

Three recent competing views on the detailed mechanism of the reaction use different means to resolve a central difficulty. A benzene ring is bulky and 'thick', so that as long as the N—N and N—C bonds of hydrazobenzene remain normal covalencies, the 4- and 4′-carbon atoms cannot approach within the length of even a quite incipient covalency.

Dewar envisages [1] a heterolytic fission of monoprotonated hydrazobenzene:

$$Ph\overset{+}{N}H_2NHPh \longrightarrow PhNH_2 + H\overset{-}{N}\text{—}C_6H_4(2^+)$$

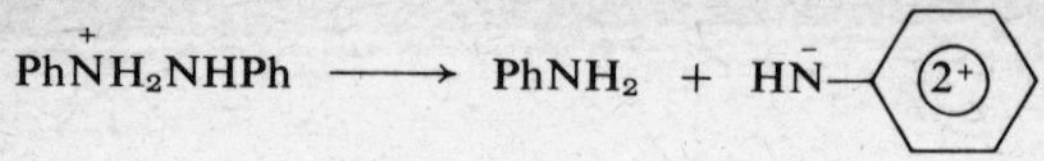

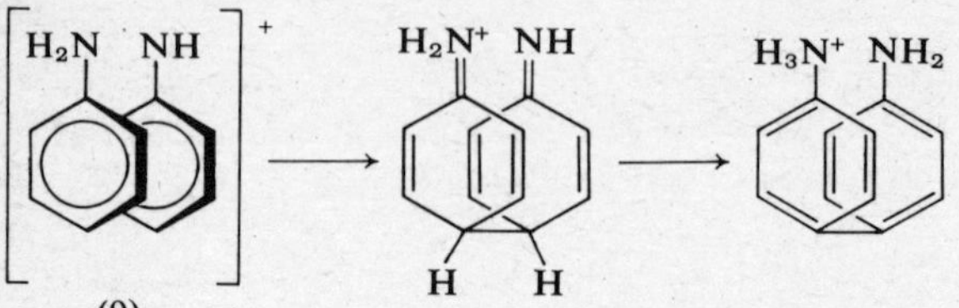

(9)

The second product (shown in one of several possible formulations) is well designed to enter promptly (in advance of kinetic freedom) into a stable π-complex with aniline (9). In a separate step, bond formation between positions 4 and 4′, accompanied by dissolution of the π-linkage, gives a tautomer of protonated benzidine.

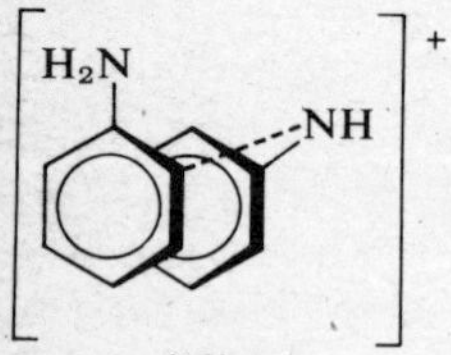

(10) 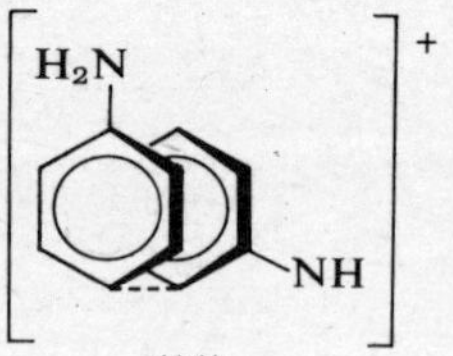 (11) 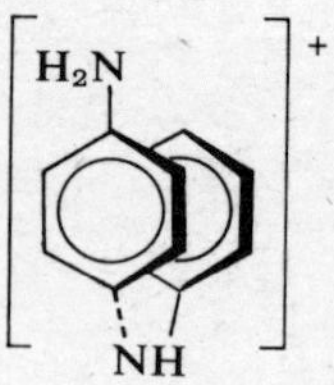 (12)

By relative rotation of the parallel rings in (9), we arrive at the isomers (10–12). Since the orbitals involved in π-complex formation are mainly concentrated on nitrogen and the *o*- and *p*-carbon atoms, the configurations (9) and (11) will predominate. The conversion of these complexes into the final products, as indicated by the dotted lines, is comparable with an electrophilic substitution into aniline, so that (9) gives benzidine (main product), (11) yields diphenyline (major by-product), and (10) and (12) give the limited yield of semidines. In the naphthalene series, maximal stability of the π-complex will be reached if all four nuclei are kept face-to-face, corresponding to formula (9), and leading to benzidine and *ortho*-benzidine but no diphenyline.

Večeřa modernised [15] Tichwinsky's suggestion, postulating a radical dissociation of diprotonated hydrazobenzene:

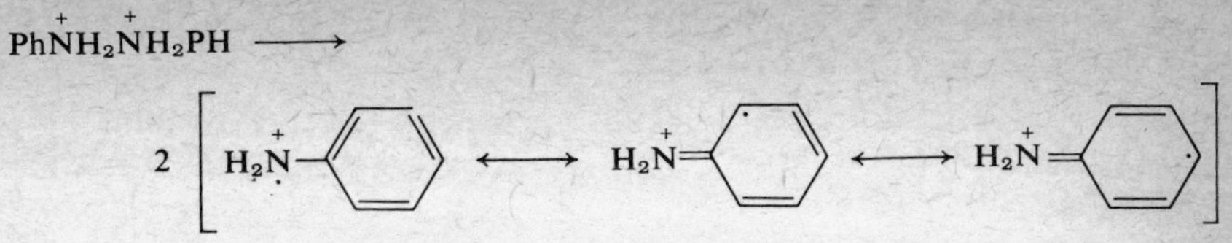

$Ph\overset{+}{N}H_2\overset{+}{N}H_2Ph \longrightarrow$

$$2\left[H_2\dot{\overset{+}{N}}-C_6H_5 \longleftrightarrow H_2\overset{+}{N}=C_6H_5^{\cdot} \longleftrightarrow H_2\overset{+}{N}=C_6H_5^{\cdot}\right]$$

The radical-cations are retained in a solvent cage before recombining to give the observed products, all reasonably accommodated. When hydrazobenzene is rearranged in a solvent of low dielectric constant, the production of diphenyline rises relatively to that of benzidine. This is plausibly attributed to increased mutual repulsion of cationic nitrogen atoms in the radical-pair [compare (9) and (11) above]. There might be some doubt as to the ability of a solvent cage to restrain so obstreperous an exhibit as a pair of cations. The theory does little to explain substituent effects, and it has been opposed on kinetic grounds.

The recent work [16] of Ingold, Banfield, and their collaborators has been based on a different line of thought. Kinetic studies, and the fact that rearrangement goes faster with D^+ in D_2O than with H^+ in solvent H_2O, indicate that mono- (and di-) protonation precede the rearrangement proper. Replacement by deuterium of the aryl protons to be displaced in the rearrangement does not affect the overall rate; this displacement therefore follows the actual rearrangement.

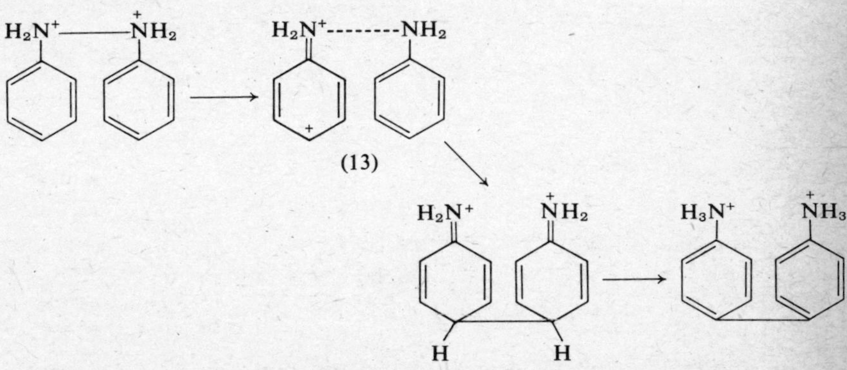

H_2N^+———$\overset{+}{N}H_2$ ⟶ H_2N^+--------NH_2

(13)

↘ H_2N^+ $\overset{+}{N}H_2$ (H, H) ⟶ H_3N^+ $\overset{+}{N}H_3$

Ingold postulated a substantial degree of heterolysis of the nitrogen-nitrogen bond in the protonated hydrazo-compound, leading to a state (13) in which the bond, still strong enough to hold the molecule together, can undergo considerable stretching and still more bending. The charge distribution in the left-hand nucleus of (13) is, by electrostatic necessity, as shown; in the other nucleus it is essentially that of an aniline molecule. With the complaisant adaptability of the partially heterolysed bond, the preferred configuration for the transition state will be (14) with the nuclear positive charge of the left-hand ring opposite the centroid of the partial negative charges of the other. This is in favour of the production of benzidine or diphenyline as against *ortho*-benzidine.

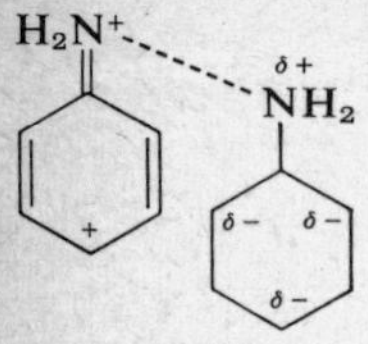

(14) 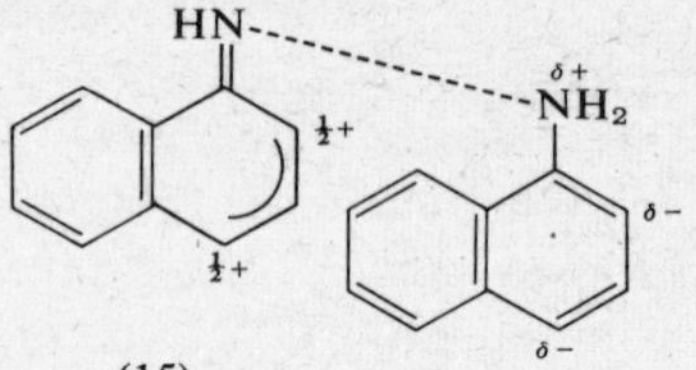 (15)

N.B. The opposite charges hold the naphthalene units in parallel planes in an eclipsed arrangement (no lateral displacement)

The corresponding picture for hydrazonaphthalene (one-proton kinetics) is (15), and the significant products are the benzidine and *ortho*-benzidine. The same applies to the rearrangement of naphthylphenylhydrazines, which also give much *ortho*-benzidine and no diphenyline.

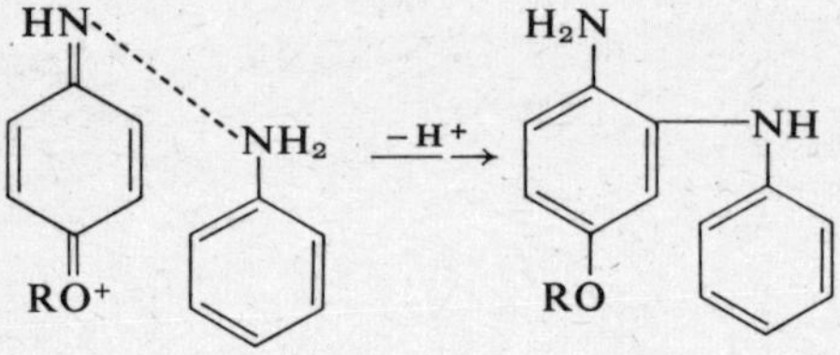

(16)

With an electron-donating *p*-substituent and one-proton kinetics, we have the situation (16) in which the right-hand ring has slipped still further downwards than in (14) and the product is the *ortho*-semidine.

The rearrangement is, comprehensibly, promoted by electron-donating substituents. It shows a strong positive salt effect, indicating a transition state still more polar than that which precedes it (telling against Večeřa's theory).

In a case where mixed kinetics are observed, we have on Ingold's theory two parallel reactions, in one of which one proton, and in the other two protons, take part in the transition complex. Increase of hydrogen-ion concentration will favour the latter mechanism, leading to a closer approximation to two-proton kinetics; this is observed in several cases. On the Dewar theory, we have a one-proton process leading reversibly to the π-complex, followed by reaction with a second proton; the overall process will be second-order in hydrogen-ion concentration if the second stage is rate-determining, a situation which will not be promoted by increase in $[H^+]$.

Similar rearrangements can be effected in the absence of acid [17], often at a higher temperature. The relation of these 'no-proton' changes to the acid-promoted changes of hydrazobenzenes is still under investigation.

Migrations from Oxygen to an Aromatic Nucleus

Such shifts have been less extensively studied than migrations from nitrogen.

Alkyl aryl ethers are converted into *o*- and *p*-alkyl-phenols by a variety of acid reagents, e.g., aluminium chloride, antimony trichloride, dilute sulphuric acid [18]. As in the case of the Hofmann–Martius rearrangement of amines, a migrating-group can be transformed (n-butyl to sec. butyl; isobutyl to t-butyl). Kinetic studies indicate simultaneous intra- and inter-molecular reaction. Similar rearrangements can be effected without an acid reagent, evidently by a radical mechanism [19].

The Fries rearrangement. The rearrangement [20] of phenol esters (PhOCOR → *o*- or *p*-$RCOC_6H_4OH$), traditionally effected by aluminium chloride, constitutes an important synthetic method since it gives, as a rule, better yields of hydroxy-ketone than a direct Friedel–Crafts synthesis from phenol and acyl chloride.

There is a general tendency for the *o*-hydroxy-ketone to be produced at high temperatures, the *p*-isomer at low. The *p*-product can be changed into the *o*-compound by heating with aluminium chloride, but this appears to require a higher temperature than a direct *o*-isomerisation of the ester. Mixture experiments suggest an intermolecular mechanism, with separation of either RCOCl or $R\overset{+}{C}O$, but their significance is diminished by the fact that aluminium chloride also effects 'redistribution' in a mixture of esters:

$$ArOCOR + Ar'OCOR' \rightleftarrows ArOCOR' + Ar'OCOR$$

As traditionally conducted, the process is heterogeneous and thus little amenable to kinetic study; it can be effected homogeneously in nitrobenzene, with at least in some cases an increased tendency to give the *o*-product. It is a daunting consideration for the kineticist that the ester can quickly coordinate with aluminium chloride, to a not easily assessable extent, at *either* or *both* oxygen atoms. Moreover, the product may be $RCOC_6H_4OAlCl_2$, further leading to a stable chelate for the *o*-isomer. It seems, however, probable that the predominance of *o*-hydroxy-ketone at high temperature is not simply the result of thermodynamic control favouring the stable chelate.

In cold nitrobenzene, *p*-tolyl acetate does not react with aluminium chloride, but addition of thymol gives *C*-acetylated thymol. Baltzly emphasises [21] that a complex of ester and aluminium chloride can act as an acylating agent, either on a molecule of ester or on a second reactive substance if present, e.g., thymol or diphenyl ether. Substitution (e.g., bromination) in phenols or their esters, in the presence or absence of aluminium chloride, gives only *p*-products. He considers that the Fries *p*-arrangement is the result of attack of coordinated ester on a second molecule, while *o*-rearrangement is an intramolecular process having a high energy of activation, and so favoured by rise in temperature.

In summary, no final conclusion as to the nature of the Fries rearrangement is yet possible. The change can also be carried out photochemically in good yield, apparently by a radical mechanism [22].

References

1. M. J. S. Dewar in *Molecular Rearrangements*, Part 1, p. 323; ed. de Mayo, Interscience (New York), 1963.
2. K. J. P. Orton, F. G. Soper, and G. Williams, *J. Chem. Soc.*, 1928, 998.
3. G. C. Israel, A. W. N. Tuck, and F. G. Soper, *J. Chem. Soc.*, 1945, 547.
4. H. Goldschmidt and B. Bardach, *Ber.*, 1892, **25**, 1347. *Cf.* H. V. Kidd, *J. Org. Chem.*, 1937, **2**, 198.
5. W. J. Hickinbottom, *J. Chem. Soc.*, 1934, 1700.
6. P. W. Neber and H. Rauscher, *Annalen*, 1942, **550**, 182.
7. G. Steel and D. L. H. Williams, *Chem. Comm.*, 1969, 975.
8. S. Brownstein, C. A. Bunton, and E. D. Hughes, *J. Chem. Soc.*, 1958, 4354; D. V. Banthorpe, E. D. Hughes, and D. L. H. Williams, *ibid.*, 1964, 5349.
9. W. N. White and J. T. Golden, *Chem. and Ind.*, 1962, 138.
10. W. N. White, D. Lazdins, and H. S. White, *J. Amer. Chem. Soc.*, 1964, **86**, 1517.
11. Y. Okamoto and H. C. Brown, *J. Org. Chem.*, 1957, **22**, 485.
12. P. Jacobson, *Annalen*, 1922, **428**, 76.

13. D. H. Smith, J. R. Schwartz, and G. W. Wheland, *J. Amer. Chem. Soc.*, 1952, **74**, 2282.
14. C. K. Ingold, *Structure and Mechanism in Organic Chemistry*, Second Edn., Chapter XII; Cornell Univ. Press (Ithaca), 1969.
15. M. Večeřa *et al.*, *Coll. Czech. Chem. Comm.*, 1960, **25**, 1992; 1966, **31**, 3486.
16. Several papers, *J. Chem. Soc.*, 1962; see C. K. Ingold, reference 14.
17. V. O. Lukashevich and L. G. Krolik, *Doklady Akad. Nauk S.S.S.R.*, 1962, **147**, 1090.
18. M. J. S. Dewar and N. A. Puttnam, *J. Chem. Soc.*, 1959, 4080, 4086; N. M. Cullinane, R. A. Woolhouse, and G. B. Carter, *ibid.*, 1962, 2995.
19. F. M. Elkobaisi and W. J. Hickinbottom, *J. Chem. Soc.*, 1960, 1286.
20. A. H. Blatt, *Org. Reactions*, 1942, **1**, 342.
21. R. Baltzly, W. S. Ide, and A. P. Phillips, *J. Amer. Chem. Soc.*, 1955, **77**, 2522.
22. H. Kobsa, *J. Org. Chem.*, 1962, **27**, 2293.

Review Articles

General reviews

M. J. S. Dewar, reference 1.

C. K. Ingold, reference 14.

D. V. Banthorpe in *The Chemistry of the Amino Group*, pp. 585–668, ed. Patai, Interscience (New York), 1968.

Orton rearrangement

P. B. D. de la Mare and J. H. Ridd, *Aromatic Substitution—Nitration and Halogenation*, pp. 151–155; Academic Press (New York), 1959.

Benzidine rearrangement

P. Jacobson, reference 12.

E. D. Hughes and C. K. Ingold, *Quart. Rev.*, 1956, **6**, 34.

Fries rearrangement

A. H. Blatt, reference 20.

A. Gerecs in *Friedel–Crafts and Related Reactions*, vol. 3, pp. 499–533; ed. Olah, Interscience (New York), 1964.

Wallach rearrangement

E. Buncel in *Mechanisms of Molecular Rearrangements*, ed. Thyagarajan, Interscience (New York), 1968.

Index

Chapter numbers are indicated in bold type